RECENT ADVANCES IN ROBOT LEARNING

edited by

Judy A. Franklin
GTE Laboratories

Tom M. Mitchell
Carnegie Mellon University

Sebastian Thrun
Carnegie Mellon University

A Special Issue of
MACHINE LEARNING
An International Journal
Vol. 23, Nos. 2 & 3
May/June 1996

KLUWER ACADEMIC PUBLISHERS
Boston / Dordrecht / London

THE KLUWER INTERNATIONAL SERIES IN ENGINEERING AND COMPUTER SCIENCE

KNOWLEDGE REPRESENTATION, LEARNING AND EXPERT SYSTEMS

Consulting Editor

Tom Mitchell

Other books in the series:

EXPLANATION-BASED NEURAL NETWORK LEARNING: A Lifelong Learning Approach
S. Thrun
ISBN: 0-7923-9716-9
MULTISTRATEGY LEARNING, R.S. Michalski
ISBN: 0-7923-9374-0
ROBOT LEARNING, J.H. Connell, S. Mahadevan
ISBN: 0-7923-9365-1
GENERATING ABSTRACTION HIERARCHIES: An Automated Approach to Reducing Search in Planning, C. A. Knoblock
ISBN: 0-7923-9310-4
REINFORCEMENT LEARNING, T. S. Sutton
ISBN: 0-7923-9234-5
AN INTRODUCTION TO FUZZY LOGIC APPLICATIONS IN INTELLIGENT SYSTEMS, R. Yager, L. Zadeh
ISBN: 0-7923-9191-8
CONNECTIONIST APPROACHES TO LANGUAGE LEARNING, D. Touretzky
ISBN: 0-7923-9216-7
STRUCTURE LEVEL ADAPTATION FOR ARTIFICIAL NEURAL NETWORKS, T. C. Lee
ISBN: 0-7923-9151-9
COMPETITIVELY INHIBITED NEURAL NETWORKS FOR ADAPTIVE PARAMETER ESTIMATION, M. Lemmon
ISBN: 0-7923-9086-5
INCREMENTAL VERSION-SPACE MERGING: A General Framework for Concept Learning, H. Hirsh
ISBN: 0-7923-9119-5
LEARNING WITH NESTED GENERALIZED EXEMPLARS, S. L. Salzberg
ISBN: 0-7923-9110-1
KNOWLEDGE ACQUISITION: Selected Research and Commentary, S. Marcus
ISBN: 0-7923-9062-8
CHANGE OF REPRESENTATION AND INDUCTIVE BIAS, D. P. Benjamin
ISBN: 0-7923-9055-5
MACHINE LEARNING, META-REASONING AND LOGICS, P. B. Brazdil, K. Konolige
ISBN: 0-7923-9047-4
AUTOMATING KNOWLEDGE ACQUISITION FOR EXPERT SYSTEMS, S. Marcus
ISBN: 0-89838-294-7
MACHINE LEARNING OF ROBOT ASSEMBLY PLANS, A. M. Segre
ISBN: 0-89838-269-6
LEARNING FROM GOOD AND BAD DATA, P. D. Laird
ISBN: 0-89838-263-7
A CONNECTIONIST MACHINE FOR GENETIC HILLCLIMBING, D. H. Ackley
ISBN: 0-89838-236-X

Machine Learning

Volume 23, Nos. 2/3, May/June 1996

Special Issue on Robot Learning

Guest Editors: Judy A. Franklin, Tom M. Mitchell and Sebastian Thrun

Distributors for North America:
Kluwer Academic Publishers
101 Philip Drive
Assinippi Park
Norwell, Massachusetts 02061 USA

Distributors for all other countries:
Kluwer Academic Publishers Group
Distribution Centre
Post Office Box 322
3300 AH Dordrecht, THE NETHERLANDS

Library of Congress Cataloging-in-Publication Data

A C.I.P. Catalogue record for this book is available
from the Library of Congress.

Printed on acid-free paper.

Printed in the United States of America

Machine Learning, 23, 117–119 (1996)

Introduction

This special issue contains seven papers on robot learning written by leading researchers in the field. As the selection of papers illustrates, the field of robot learning is both active and diverse. A variety of machine learning methods, ranging from inductive logic programming to reinforcement learning, is being applied to many subproblems in robot perception and control, often with objectives as diverse as parameter calibration or concept formation.

While no unified robot learning framework has yet emerged to cover the variety of problems and approaches described in these papers and other publications, a clear set of shared issues underlies many robot learning problems.

- Machine learning, when applied to robotics, is situated: it is embedded into a real-world system that tightly integrates perception, decision making and execution.
- Since robot learning involves decision making, there is an inherent active learning issue.
- Robotic domains are usually complex, yet the expense of using actual robotic hardware often prohibits the collection of large amounts of training data.
- Most robotic systems are real-time systems. Decisions must be made within critical or practical time constraints.

These characteristics present challenges and constraints to the learning system. Since these characteristics are shared by other important real-world application domains, robotics is a highly attractive area for research on machine learning.

On the other hand, machine learning is also highly attractive to robotics. There is a great variety of open problems in robotics that defy a static, hand-coded solution. Real robot hardware, sensors, and environments impose intrinsic limitations on the abilities of humans to model and thus program reliable behaviors. Machine learning techniques may allow the robot to adapt to its environments, its sensors and actuators, and to acquire new skills on-the-fly without the need for further programming. As the papers here demonstrate, there are many examples in which machine learning techniques have successfully improved the competence, robustness, and efficiency of robot control programs—many of these results are hard to imagine without the use of programs that are able to learn. It is even harder to imagine a future of robotics without machine learning. Building an intelligent robot has been a fundamental goal of artificial intelligence since its earliest days. If intelligent robots are ever to become a reality, robot learning will play a role!

ROAD-MAP

The following paragraphs briefly summarize the seven papers in this issue.

- In "Real-World Robotics: Learning To Plan for Robust Execution," Bennett and DeJong introduce an approach called permissive planning, where the permissiveness of a plan is a measure of how closely the plan's preconditions must match the real-world for the plan to succeed. A combination of explanation-based learning to acquire plan schemata and a new approach to the refinement of plans is described, along with an implementation on a hardware robot arm.

- "Robot Programming by Demonstration (RPD): Supporting the Induction by Human Interaction," by Friedrich et al. combines analytical and inductive learning to generalize the notion of teaching a robot by example, using only a few examples of the proper sequence of motions. On top of this is dialog based learning, a series of questions and answers that occurs while the human is demonstrating. This additional level helps the system determine the intent of the human and narrows the hypothesis space. Implementation involves a physical robot arm.

- Chen et al. present "Performance Improvement of Robot Continuous-Path Operation through Iterative Learning Using Neural Networks." Performance improvement of continuous path operation is the control engineer's approach to solving the problem of explicitly teaching every move to a robot. This paper has a tutorial nature in that concepts from engineering robotics, such as closed-loop stability and PID control are clearly described.

- "Learning Controllers for Industrial Robots," by Baroglio et al. is a summarization and comparison of a number of machine learning techniques designed for nonlinear systems. Included are Multilayer Perceptrons, Radial Basis Functions, and Fuzzy Controllers. The comparison takes the form of algorithmic and empirical analysis. This sets the stage for the description of two original integrated learning algorithms.

- In "Active Learning for Vision-Based Robot Grasping," Salganicoff et al. employ a new integrated learning algorithm, IE-ID3, to give active-learning ability to a robot arm equipped with a vision system. The task is to choose appropriate grasping approaches in order to pick up various objects. Two important objectives are for the algorithm to produce real-valued actions and for learning to occur quickly.

- "Purposive Behavior Acquisition for a Real Robot by Vision-Based Reinforcement Learning," by Asada et al. describes a mobile robot that learns to shoot a ball into a goal, using Q-learning. Environmental information is given only by the visual image. In addition to Q-learning, the authors employ a learning schedule called Learning from Easy Missions. This paradigm includes an algorithmic decision maker for shifting to more difficult tasks.

- The final paper, "Learning Concepts from Sensor Data of a Mobile Robot" by Klingspor et al., also uses a mobile robot as a platform to explore the use of machine learning in the link between low-level representations of sensing and action and the high-level representation of planning. The applications addressed involve robots that are not completely autonomous, but interact with human users. The machine learning technique employed is inductive logic programming with some modifications such as bounding the number of rules and predicates as well as splitting the overall task into several learning steps.

The editors would like to thank the authors who submitted papers to this issue, and especially like to thank the reviewers, all of whom provided extensive feedback and comments to us and the authors.

Judy A. Franklin, GTE
Tom M. Mitchell, CMU
Sebastian Thrun, CMU and University of Bonn

Machine Learning, 23, 121–161 (1996)

Real-World Robotics: Learning to Plan for Robust Execution

SCOTT W. BENNETT bennett@sra.com
Systems Research and Applications Corporation, 2000 15th Street North, Arlington, VA 22201

GERALD F. DEJONG dejong@cs.uiuc.edu
Beckman Institute, University of Illinois, 405 North Mathews Avenue, Urbana, IL 61801

Editors: Judy A. Franklin, Tom M. Mitchell, and Sebastian Thrun

Abstract. In executing classical plans in the real world, small discrepancies between a planner's internal representations and the real world are unavoidable. These can conspire to cause real-world failures even though the planner is sound and, therefore, "proves" that a sequence of actions achieves the goal. Permissive planning, a machine learning extension to classical planning, is one response to this difficulty. This paper describes the permissive planning approach and presents GRASPER, a permissive planning robotic system that learns to robustly pick up novel objects.

Keywords: machine learning, robotics, uncertainty, planning

1. Introduction

For some time technical difficulties have prevented the field of planning from delivering on its early promise. Increasingly it has become accepted that domain-independent classical planning, in which one finds a sequence of actions which provably achieves a goal when applied to a problem's initial state is at best quixotic. The death knell for the approach was sounded by Chapman (Chapman, 1987) who showed that with some assumptions (which are rather reasonable in real-world domains), the general classical planning problem is prohibitively difficult. One new approach, known generally as *reactivity*, has made great strides to circumvent the obstacles of classical planning (Agre & Chapman, 1987, Firby, 1987, Schoppers, 1987, Suchman, 1987). Classical planning demands that the planner know and model the effects of its actions. For any sequence of actions a classical planner must be able to judge accurately their cumulative changes to a state. This ability to project states through operators may seem modest at first but the failure of classical planning can be viewed as traceable to this task. A central tenet of pure reactive activity is to do no projection. The "planner" (or better the "agent") makes no attempt to anticipate how the world will look after an action is executed. Instead, an action is selected based entirely upon the agent's sensor-supplied knowledge of the current world state. After the execution of an action, the world changes in some (possibly complex) way. The next action is selected in the same manner, based upon the updated sensor values. In a way, a purely reactive system employs the world itself to model the effects of actions at the expense of *gedanken* backtracking. The method of selecting actions, and the demands such a mechanism places upon its sensors, are the subject of much current research

(Agre & Chapman, 1987, Hammond, Converse, Marks, 1990, Brooks, 1987). In any case, the resulting system relies on its sensor abilities to drive action selection.

Models of reactivity, for all their recent popularity, are only one approach to the problems faced by classical planning. In this paper we introduce another. In some ways it is the dual of the reactive approach. Our approach, called *permissive* planning, like the reactive approach, gives up the notion of a provably correct plan. However, the concept of projection remains. Indeed, it is, if anything, more central than before.

In most real-world domains it is impossible to describe the world correctly and completely. It follows that internal system representations of the world must, at best, be approximate. Such approximate representations of world attributes are called *data approximations*. They may arise from imperfect sensors, or incomplete inferencing. We introduce the concept of *permissiveness* of a plan as a measure of how faithfully the plan's preconditions must reflect the real world in order for the plan to accomplish its goals. One plan is more *permissive* than another if its representations can be more approximate while continuing to adequately achieve its goals. We do not propose to quantify this notion of permissiveness. Instead, we employ a machine learning approach which enhances permissiveness of acquired planning concepts.

The approach involves acquiring and refining generalized plan schemata which achieve often-occurring general goals and sub-goals. Acquisition is through rather standard explanation-based learning (Serge, 1988, Mitchell, Keller & Kedar-Cabelli, 1986, DeJong & Mooney, 1986, Mitchell, Mahadevan & Steinberg, 1985). However, the refinement process is unique.

To drive refinement, the system constantly monitors its sensors during plan execution. When sensor readings fall outside of anticipated bounds, execution ceases and the plan is judged to have failed. The failure must be due to a data approximation; if there were no mismatch between internal representations and the real world, the plan would have the classical planning property of provable correctness. The plan's failure is diagnosed. Ideally, only a small subset of the system's data approximations could underlie the monitored observations. The system conjectures which of its internal expressions might account for the observations. Next, the system uses qualitative knowledge of the plan's constituent operators. The small conjectured error is symbolically propagated through the plan to plan parameters. The plan parameters are adjusted so as to make the planning schema less sensitive to the diagnosed discrepancy with the world. If the process is successful, the refined schema is uniformly more permissive than the original, which it replaces. Thus, through interactions with the world, the system's library of planning schemata becomes increasingly permissive, reflecting a tolerance of the particular discrepancies that the training problems illustrate. This, in turn, results in a more reliable projection process. Notice that there is no improvement of the projection process at the level of individual operators. Performance improvement comes at the level of plan schemata whose parameters are adjusted to make them more tolerant of real-world uncertainties in conceptually similar future problems. Adjustment is neither purely analytical nor purely empirical. Improvement is achieved through an interaction between qualitative background knowledge and empirical evidence derived from the particular real-world problems encountered.

The notion of permissive planning is not tied to any particular domain. It is a domain-independent notion that is, nonetheless, not universally applicable. There are characteristics of domains, and problem distributions within domains, that indicate or counter-indicate the use of permissive planning. Later, the requirements of permissive planning will be made more formal. Here we give an intuitive account of characteristics needed to support permissive planning. An application that does not respect these characteristics is unlikely to benefit from permissive planning.

For permissive planning to help, internal representations must be reasonable approximations to the world. By this we mean that there must be some metric for representational faithfulness, and that along this metric, large deviations of the world from the system's internal representations are less likely than small deviations.

Second, some planning choices must be subject to continuous real-valued constraints or preferences. These constraints and preferences are called *parameters* of the plan schema. They might be thresholds indicating bounds on acceptable values, or actual real-valued arguments to domain operators. These parameters are tuned to achieve permissiveness.

Finally, the planner must be supplied with information on how each operator's preconditions and arguments *qualitatively* change its effects. This information is used to regress the diagnosed out-of-bounds approximations of failed plans through the planning structure to parameters. Such propagation determines how parameters may be adjusted so as to decrease the likelihood of similar future failures.

Clearly, many domains do not respect these constraints. However, robotic manipulation domains form an important class in which the above characteristics are naturally enforced. Consider the data approximation constraint. A typical expression in a robotics domain may refer to real-world measurements. Object positions and dimensions, for example, require some representation for metric quantities. An example might be something like **(HEIGHT-IN-INCHES BLOCK3 2.2)**. Such an expression is naturally interpreted as an approximation to the world. Indeed, expressions such as this one are useless under a standard semantics. The conditions of truth require that the height of the world object denoted by **BLOCK3** be exactly 2.2 inches. Technically, no deviation whatsoever is permitted. If the height of **BLOCK3** is off by only 10^{-40} inches, the expression is false – just as false as if it were off by 5 inches or 50 inches. Clearly, such an interpretation cannot be tolerated; the required accuracy is beyond the numerical representational capabilities of most computers. Another nail is driven into the coffin for standard semantics by real-world constraints. Actual surfaces are not perfectly smooth. Since the top and bottom of **BLOCK3** most likely vary by more than 10^{-40} inches, the "height" of a real-world object, using a standard semantics, is not even a well-defined relation. In fact, no working system adopts a strict standard semantic interpretation for expressions such as the one above. There are several alternatives which will be discussed later. For now, it is sufficient to notice that expressions such as **(HEIGHT-IN-INCHES BLOCK3 2.2)** are extremely common in robotic domains and can be easily interpreted as satisfying our informal definition of an approximation: the metric for faithfulness is the real-valued height measure, and, presumably, if a reasonable system describes the world using the expression **(HEIGHT-IN-INCHES BLOCK3 2.2)** it is more likely the case that **BLOCK3** is 2.2001 inches high than 7.2 inches high. It is essential that the expression not saddle the system with the claim that **BLOCK3** is precisely 2.2 inches high.

The second condition for permissive planning requires that continuous real-valued parameters exist in the system's general plans. Geometric considerations in robotic manipulation domains insure that this condition is met. As an example, consider a robot manipulator that wishes to move its arm past **BLOCK3**, which in turn rests on the table. Some path must be adopted for the move. From the geometrical constraints there is a minimum height threshold for the path. Since the arm must not collide with anything (in particular with **BLOCK3**), it must be raised more than 2.2 inches above the table. This height threshold is one of the plan parameters. Any value greater than 2.2 inches would seem to be an adequate bound on the parameter for the specific plan. However, it is easy to see that if 2.2 inches *is* adequate, so is 2.3 inches, or 5.0 inches, etc. Thus, the plan supports the parameter as a continuous real-valued quantity. Notice, that once the specific plan of reaching over **BLOCK3** is generalized by EBL, the resulting plan schema parameterizes the world object **BLOCK3** to some variable, say, *?x* and the value 2.2 to *?y* where **(HEIGHT-IN-INCHES** *?x ?y*) is believed, and the threshold parameter to *?z* where *?z* is equivalent to (+ *?y* ϵ) for the tight bound, or equivalent to (+ *?y* ϵ 0.1), for the second bound above, or equivalent to (+ *?y* ϵ 2.8), for the third bound above. The value of ϵ insures that the bound is not equaled and can be made arbitrarily small in a perfect world. As will become clear, in permissive planning, ϵ may be set identically to zero or left out entirely.

The final condition for permissive planning requires qualitative information specifying how the effects of domain operators relate to their preconditions and arguments.[1] This constraint, too, can be naturally supported in robotic manipulation domains. Consider again the plan of moving the robot arm past **BLOCK3**. The plan involves moving the arm vertically to the height *?z* and then moving horizontally past the obstacle. The required qualitative information is that the height of the robot arm (the effect of **MOVE-VERTICALLY**) increases as its argument increases and decreases as its argument decreases. With this rather simple information the generalized plan schema for moving over an obstacle can be successfully tuned resulting in a more permissive plan schema.

To illustrate, suppose the plan schema for moving past an obstacle is acquired from a successful specific example of moving past **BLOCK3**. To have been successful, reality must not have interfered with the system's approximations. Suppose **BLOCK3** is in reality 2.14962 inches high. The naive specific plan simply moves the arm to the minimum threshold (2.2 inches, the believed height of **BLOCK3**) successfully missing the obstacle by about a twentieth of an inch. Since it was successful, a schema might be generalized for later use. This involves variablizing the constants in standard EGGS (DeJong & Mooney, 1986) or EBG (Mitchell, Keller & Kedar-Cabelli, 1986) fashion. Suppose now that we wish to reach past another obstacle, say **BLOCK7** which we believe is 5.5 inches high, but is in reality 5.6 inches high. The plan fails when an unexpected force is encountered during the horizontal move action. The failure must be diagnosed to a set of likely offending approximations. For the example, we will assume that diagnosis uniquely identifies **(HEIGHT-IN-INCHES BLOCK7 5.5)** as the offending representation, conjecturing that its height is greater than represented. This might be done from the fact that the unexpected force is perpendicular to the direction of motion, and **BLOCK7** is believed to be the only object in the vicinity of the arm at the time of collision. The result of the diagnosis is a symbolic expression Δ representing that the block is too high for the plan to succeed. Let $\mathbf{P}(\boldsymbol{\Delta})$ represent the probability that the general obstacle is too high. We do not attempt to assign an actual numerical probability to this expression; it is treated purely symbolically. If it is decided that the system must not tolerate the failure, then $\mathbf{P}(\boldsymbol{\Delta})$, whatever its value, is greater than the allowable threshold. The generalized plan must be adjusted to reduce $\mathbf{P}(\boldsymbol{\Delta})$, or the planning concept must be discarded. The next step is to qualitatively regress $\mathbf{P}(\boldsymbol{\Delta})$ through the plan to parameters that unambiguously, in a qualitative sense, reduce $\mathbf{P}(\boldsymbol{\Delta})$. If no such parameters can be found, the general planning concept must be discarded. According to the plan, the parameter *?z* above is identified as one whose increase uniformly reduces $\mathbf{P}(\boldsymbol{\Delta})$. That is, as one moves "excessively" higher over an obstacle of uncertain height, one is less likely to collide with it. If there is no conflicting bound on *?z* it is adjusted upward, and the refined plan replaces the original. Thus, in the future similar planning episodes the robot system will raise its hand higher than it believes necessary when passing over obstacles. The permissiveness of the plan has been increased. Increasing *?z* to decrease $\mathbf{P}(\boldsymbol{\Delta})$ may increase the probability of some *other* failure not encountered in the refining experience. Managing and exploiting tradeoffs in the context of a particular problem distributions is what permissive planning is all about.

From a different point of view, permissive planning amounts to blaming the plan for execution failures, even when in reality the representations, not the plan, are at fault. This is a novel approach to planning which results in a different, rather strange semantics for the system's representations. Current research includes working out a more formal account of the semantics for representations in permissive plans. Straightforward interpretations of the expressions as probabilistic seem not to be sufficient. Nor are interpretations that view the expressions as fuzzy or as having uncertainty or error bounds. The difficulty lies in an inability to interpret an expression in isolation. An expression "correctly" describes a world if it adequately supports the permissive plans that make use of it. Thus, an expression cannot be interpreted as true or not true of a world without knowing the expression's context including the system's planning schemata, their permissiveness', and the other representations that are believed.

We now return to the popular alternatives to uncertainty in robotics, alternatives to a standard semantics for expressions such as **(HEIGHT-IN-INCHES BLOCK3 2.2)**. Inability to precisely describe the world has always been a difficult problem in robotics domains. There are three common approaches. The first most obvious approach is to select representations that are, in a sense, good enough. This is the traditional AI notion of a micro-world. In the most straightforward version, the system implementor takes on the responsibility for insuring that no problems will result from necessarily imprecise descriptions of the domain. In general, this requires the implementor to characterize in some detail all of the future processing that will be expected of the system. Often he must anticipate all of the planning examples that the system will be asked to solve. If the physical robot system is not up to the accuracy that the examples require, the implementor must build a better vision system or purchase a more precise, more reproducible robot manipulator. This approach has enjoyed frightening popularity. While it is most often used in systems which research phenomena other than uncertainty, the implementors seldom more than tacitly acknowledge the semantic implications. When employed by practical systems the approach results in an ultimately-doomed quest for increasingly exacting (and expensive) hardware. The great irony of industrial automation, where this approach is nearly universal, is that the mechanical positioning capabilities of robots must far exceed the humans that they replace. The characteristic brittleness and inflexibility of industrial robotics is a consequence of the semantically-rooted constraint that the implementor anticipate all future applications

The second approach involves monitoring and/or manipulating explicit representations of uncertainty (Zadeh, 1965, Brooks, 1982, Lozano-Perez, MAson & Taylor, 1984, Erdmann, 1986, Davis, 1986, Hutchinson & Kak, 1990). In a pure and unsimplified form this approach adequately models any knowable theory of uncertainties; it preserves the classical planning ideal of a provably correct plan. Unfortunately, a high computational price is incurred. A general ability to project states including objects with explicit general error bounds is necessarily no less difficult than if the objects are known precisely. Many systems incorporate simplifications, typically assuming that uncertainties are constant, independent of context, or otherwise constrained. This buys some efficiency at the price of generality. But the efficiency is never greater than if the objects had zero uncertainty so that the Chapman intractability results still apply.

Finally, spatial uncertainties in robotics may be dealt with by guaranteeing a conservative approach to inferencing. In a sense, a kind of worst (or sufficiently bad) case representation is assumed for objects. The approach, which seems only used for problems of path planning, includes techniques like quantizing the space (Wong & Fu, 1985, Zhu & Latombe, 1990, Malkin & Addanki, 1990) and imagining a repulsive potential field around obstacles (Khatib, 1986, Hwang, 1988). Interestingly, the approach can be more efficient than the zero uncertainty case. Since object boundaries are not guaranteed to be the tightest possible, they can be selected to be both conservative and simplifying. This benefit does not come for free. In different guises completeness or correctness can be sacrificed. In a sense, this is the closest of the popular approaches to the research reported in this paper. In a sense we also adopt a conservative representation, although the uncertainty tolerance is due to plan characteristics rather than explicit representations. This shift supports a context-sensitive conservatism which supports reasoning about general manipulation problems rather than only path planning in a static world.

2. Learning Increasingly Permissive Plans

This section defines a framework for permissive planning and describes an algorithm which, based on execution failures, incrementally increases plan permissiveness so as to decrease the likelihood of those failures. First, data approximations are defined. They provide the internal representations employed during planning. Because data approximations are approximations to the world, plan execution failures which may result can be linked to some set of these data approximations. Next, we define plan parameters. These parameters may be tuned to increase the permissiveness of the plan. The remainder of this section discusses the plan generation, execution, and refinement process for permissive plans.

2.1. *Data Approximations*

Data approximations are representations for approximate continuously valued data about the state of the world. They can either be *external* or *internal*. External data approximations are used to represent the uncertainty of data in the world. Internal data approximations are used to simplify complex sets of data to make reasoning more tractable. First, let us consider external data approximations.

2.1.1. *External Data Approximations*

An external data approximation involves a set of quantities for which the system is given approximate values typically via imperfect sensors. Let Q_E be a vector $\{q_1, q_2, q_3, ..., q_n\}$of quantity variables and A_E be a vector of their corresponding approximate values $\{a_1, a_2, a_3, ..., a_n\}$. Every q_i exists along a continuous dimension $D(q_i)$. We adopt the following model for external data approximations. For each quantity q_i there is a measurable value knowable to the system denoted $M_V(q_i)$. There is also an actual world value not knowable to the system denoted $A_V(q_i)$. For a sensor to yield a valid approximation we require that $P(M_V(q_i) = a_i | A_V(q_i) = t)$ is monotonic in $|t - a_i|$. In other words, given the actual value of t of a sensed quantity $q_i(A_V(q_i) = t)$, the likelihood of measuring value a_1 is greater than measuring value a_2 in just those cases that a_1 is closer to t than is a_2. More precisely:

$$\forall_{i=1}^{n} P(M_V(q_i) = a_1) > P(M_V(q_i) = a_2) \iff |a_1 - t| < |a_2 - t| \tag{1}$$

In the case of external data approximations, the value vector A_E is the best information the system has about the values of the quantity variables Q_E. The only way to improve this information is to interact with the world. For purposes of planning with the data represented by the approximations, the system behaves as if $Q_E = A_E$. The qualitative definition of a data approximation is never employed during planning, only when analyzing failures.

For instance, suppose one is using a tape measure to measure the length of a piece of wood. The reading obtained fits the criteria for an external data approximation. The quantity q_i being measured is the length of the wood and a_i is its measured approximate value. The quantity q_i exists along the continuous dimension $D(q_i)$ of length. It is reasonable to believe that the error between the measured and *true* value is likely to be small. The only way better information can be obtained is through further interaction with the world: obtaining and using a better measuring device, measuring more times and averaging the results, etc...

In robotics, external data approximations can be used to represent values read from sensors, which are inherently uncertain. For instance, the position of a block, as sensed by a visual system, would be represented with an external data approximation.

2.1.2. *Internal Data Approximations*

With an internal data approximation, the system chooses the values A_I of the quantity variables Q_I with a data approximation procedure. This is often motivated by the need to simplify the representation so reasoning can be performed more efficiently. Internal data approximations can be adjusted through the system's reasoning alone. One can think of internal data approximations as being like external data approximations except that the actual value of the variable represents the optimal setting as defined by a system utility function.[2] The measured value for a quantity corresponds to the value chosen by the system in an attempt to maximize utility.

Recall our earlier tape measure example. Suppose that we would like to use the measurements we are making to decide how much wood is needed for a project. We might consider rounding up to the nearest inch in making the measurements. This makes it easier to compute the wood needed at the cost of possibly requesting too much. On top of the external data approximation, due to the uncertainty of the measurement, we have added an additional internal approximation to permit easier planning of the wood requirements.

In robotics, geometric object models are examples of internal data approximations. A simplified geometric representation can be far more efficient to reason about than the complex raw data returned by a vision system. However, in seeking a simplified representation, accuracy is sacrificed. The system has thus introduced further uncertainty.

2.2. Plan Parameters

Permissive plans employ continuous numeric parameters which can be tuned to affect plan permissiveness. It is important that these parameters depend on the context in which the plan is applied. Suppose we have a parameter which specifies the height which a manipulator must be above the workspace to safely navigate without collisions. The possible settings for this parameter are a function of the highest object in the workspace. Therefore, it depends on context. This means a generalized plan schemata must choose its parameter values at plan application time.

Parameters are chosen based on a *quality* function maintained in the schemata which indicates the current "best" parameter choices for the context in which the schemata is being applied. The quality function is represented by a collection of preferences which qualitatively describe the behavior of the function over a set of intervals.

Let Q_P be the set of plan parameters and A_P be the set of their respective values. Every $q_i \in Q_P$ is defined along a continuous dimension $D(q_i)$. Let there be a low bound $L_{low}(D(q_i), C)$ and a high bound $L_{high}(D(q_i), C)$ on the values which q_i may assume along dimension $D(q_i)$ in context C. A *context* is a partial world state specification. Figure 1 gives a pictorial representation for the dimension $D(q_i)$. The plan's parameter values A_P must also be dependent on context. The system chooses A_P based on a set of preferences $P(q_i, C)$ for each q_i in context C.

A preference $p \in P(q_i, C)$ is a 3-tuple $< V_{low}(C), V_{high}(C), r >$ where $V_{low}(C)$ and $V_{high}(C)$ form an interval dependent on context, and $r \in$\{*increasing, decreasing, constant*\} is a relation describing the behavior of a *quality* function $F_{Q,q_i,C}$ in that interval. Additionally:

$$L_{low}(D(q_i), C) \leq V_{low}(C) \leq V_{high}(C) \leq L_{high}(D(q_i), C) \qquad (2)$$

$$P(q_i, C) \; is \; inconsistent \iff \exists p_i, p_j \; such \; that$$

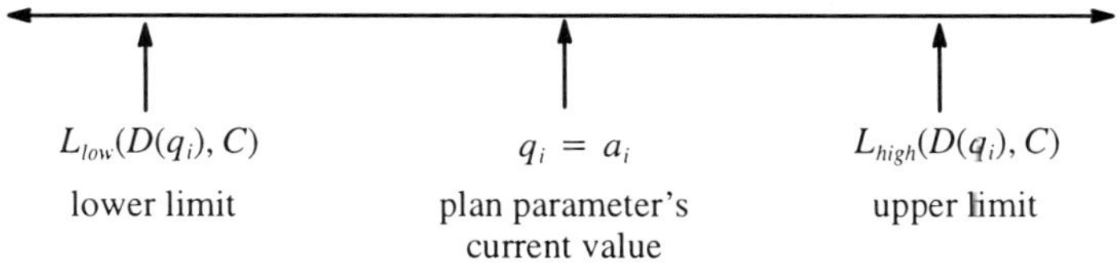

Figure 1. A Graphic Representation of Dimension $D(q_i)$

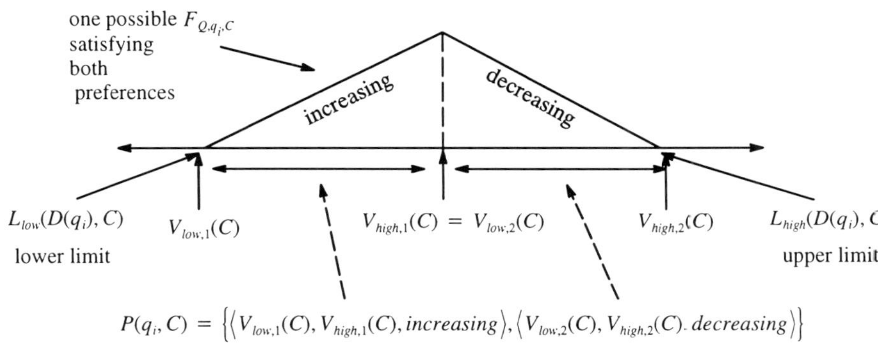

Figure 2. A Graphic Representation of Two Consistent Preferences Comprising A Quality Function

$$
\begin{gathered}
p_i = < V_{il}(C), V_{ih}(C), r_i > \in P(q_i, C) \wedge \\
p_j = < V_{jl}(C), V_{jh}(C), r_j > \in P(q_i, C) \wedge \\
r_i \neq r_j \wedge \\
\left(\begin{array}{c} V_{il}(C) < V_{jl}(C) < V_{ih}(C) \vee \\ V_{il}(C) < V_{jh}(C) < V_{ih}(C) \end{array} \right)
\end{gathered}
$$

Figure 2 shows two consistent preferences comprising a quality function. An incompletely specified preference is one which has an undetermined value for $V_{low}(C)$ or $V_{high}(C)$. If incompletely specified preferences are used, as they are in our GRASPER implementation discussed in Section 3, the system must provide a procedure for assigning values to the undetermined $V_{low}(C)$'s and $V_{high}(C)$'s. The procedure should attempt to find assignments such that the resulting set of preferences is consistent as defined above.

The overall quality function F_Q determines the choice made by the plan for A_P. $F_{Q,q_i,C}$ is only specified in terms of the qualitative behavior of its regions as defined by the preferences $P(q_i, C)$. A plan parameter q_i may not be able to take on all values a_i, where $L_{low}(D(q_i), C) \leq a_i \leq L_{high}(D(q_i), C)$, in a given context C. We therefore represent the set of values q_i can take on in context C as $R(q_i, C)$. Let $MAX(F_{Q,q_i,C}, X)$ be a set of potential maxima for the function $F_{Q,q_i,C}$ applied to the set of values X. For $x \in X$,

$L_{low}(D(q_i, C)) \leq x \leq L_{high}(D(q_i), C)$ because the function is only defined over that interval. With only qualitative information about the quality function only potential maxima can be identified. The plan must choose a value for $a_i \in MAX(F_{Q,q_i,C}, X)$. Several strategies are possible. The GRASPER implementation chooses the value closest to the current setting for q_i. In general, the plan chooses a value for one or more parameters $q_i \in MAX(F_{Q,q_i,C}, R(q_i, C))$.

We assume limited interactions between plan parameters. L_{low}, L_{high}, and all V_{low}'s and V_{high}'s are dependent on context and may be a function of other plan parameters. We enforce that they must maintain their relative ordering under any possible valuation of the other plan parameters. In general: given an ordered, consistent set of preferences $\{P_1(q_i, C), P_2(q_i, C), ..., P_n(q_i, C)\}$ where each
$P_j(q_i, C) = < V_{low,j}(C), V_{high,j}(C), rel_j >$ then:

$$\forall C[(k < l) \Rightarrow [\ L_{low}(D(q_i), C) \leq V_{low,k}(C) \leq V_{high,k}(C) \leq V_{low,l}(C) \leq V_{high,l}(C) \leq L_{high}(D(q_i), C)]] \quad (3)$$

2.3. *Planning with Data Approximations*

Constructing a plan for a specific goal making use of data approximations is the same as with traditional planning techniques. The data approximations permit a simplified model of the world, simplifying the reasoning involved. No explicit reasoning about the fact that data approximations were employed takes place during plan construction or application. Generalized plan schemata can be constructed using the EGGS algorithm (Mooney & Bennett, 1986). The choice of parameter values for the generalized plan schemata, however, will remain dependent on the context in which the generalized plan is applied.

The price for approximation of the world is the increased potential for failure due to discrepancies. However, the strength of the permissive planning approach is in the ability to tune plan parameters to succeed in spite of these discrepancies. To achieve this we require that the system have an ability to recognize failures and to tune plan parameters so as to reduce the likelihood of encountered failures. The next two subsections discuss these requirements. First, we define expectations which are constructed during the planning process and allow failures to be recognized during execution. Next, we discuss how plans are refined so as to become increasingly permissive.

2.4. *Expectations*

We are interested in improving plan performance with respect to the real world. Failures occur when feedback from the real world contradicts the prediction of the simplified model which employs data approximations. Expectations are predictions of the simplified model for an action in terms of conditions which can be efficiently monitored in the world. In order to detect execution failures, all actions carried out in the world must be monitored. The important parts of a monitor are illustrated in Figure 3. First, the monitor must specify

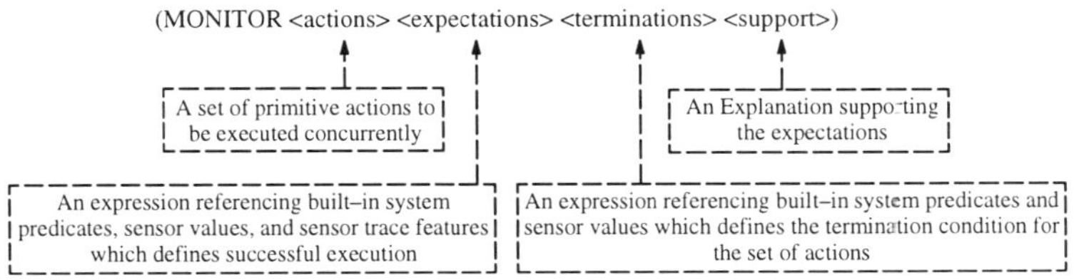

Figure 3. Syntax for Monitored Actions

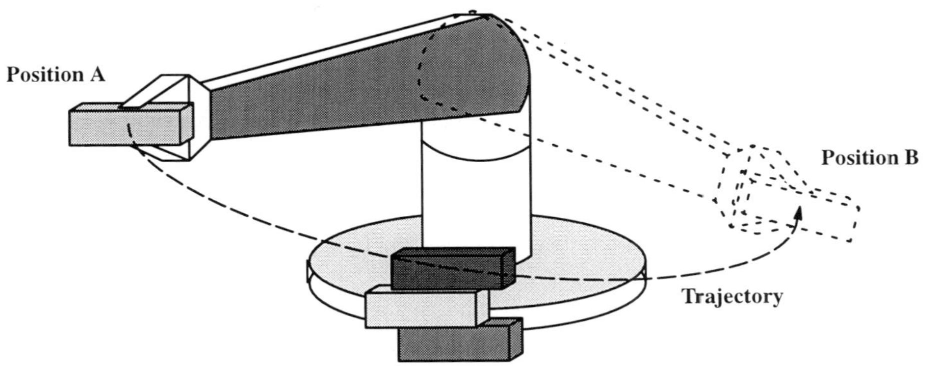

Figure 4. An Action for which Expectations Are Specified

the set of concurrent primitive actions to be carried out. Next, a set of constraints on the readings for sensors must be provided which define expected operation of the actions. A set of terminations is also provided in terms of sensor values which define termination of the actions. If the terminations are satisfied and none of the expectations were violated, the actions have succeeded. Lastly, there must exist a justification for the expectations in terms of the simplified model. As part of the planning process, all actions must have a justified set of expectations.

For instance, suppose that the robotic manipulator shown in Figure 4 is to be moved from position A to position B along some trajectory. The move is composed of a set of primitive joint moves for the manipulator which are to take place in a coordinated fashion. These form the set of concurrent primitive actions. The expectation is that the manipulator strikes nothing during the movement. This could be confirmed by force sensors on the manipulator. If none of the force sensors read sufficiently high to indicate the presence of an external force then it is assumed that the manipulator struck nothing during the move. The termination for the action set is position B as specified by the readings of a set of joint encoders on the manipulator's joints. The support for the expectations involves reasoning about the spatial occupancy of nearby objects in the simplified model. The expectation of sensing no external forces during the move could not be satisfied if the space occupied by the stack of blocks in Figure 4 intersected the space occupied by the manipulator during the trajectory.

2.5. *Refining Failing Plans for Increased Permissiveness*

Failures are defined by expectation violations. These occur when the supporting proof for an expectation is valid in the model but is contradicted by real-world experience. That difference triggers the first phase of failure recovery: generating a qualitative explanation of how to tune plan parameters to reduce the chance of the failure in the future.

2.5.1. *Qualitative Tuning Explanations*

In order to diminish the chance uncertainty-related failures, it is necessary to decide which plan parameters to tune and how to tune them. In our model, failures can always be attributed to poor data approximations. In order to devise a strategy for tuning parameters so as to decrease the likelihood of a failure, it is necessary to reason about the relationships which exist between data approximate quantities, the failing expectations, and tunable plan parameters.

We employ a qualitative model of the relationship between continuous quantities.[3] Let $Q_+(a,b)$ signify that the magnitude of quantity b positively influences the magnitude of quantity a. Similarly, $Q_-(a,b)$ means the magnitude of quantity b inversely influences the magnitude of quantity a. That is, if $a = f(b)$ the minimum we must know to create such a relation is the sign of $\frac{\partial f}{\partial b}$. If $a = f(b)$ and $\frac{\partial f}{\partial b} > 0$ then $Q_+(a,b)$ holds. If $a = f(b)$ and $\frac{\partial f}{\partial b} < 0$ then $Q_-(a,b)$ holds.

Quantitative predicates employed by the system have one of two basic intents. Either they are *calculation predicates*, whose purpose is to compute some value (e.g. a predicate for subtraction), or they are *test predicates*, which are designed to fail for certain sets of inputs (e.g. a predicate for performing a less-than comparison). There is no way to vary the probability of success of a calculation predicate since they always succeed. A test predicate's probability of success, is sensitive to the probability distribution of its argument quantities. In the diagram below, the less-than test on the right has

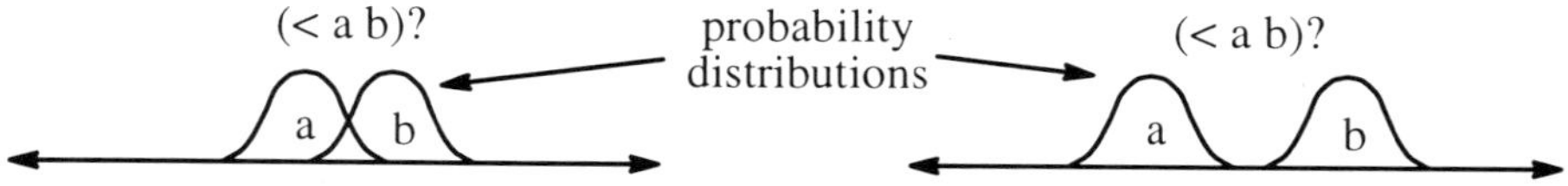

a higher probability of succeeding given the illustrated probability distributions for its arguments than the one on the left. While probability distributions are difficult to define and work with, recall the simpler qualitative view of the probability distribution defined for data approximations in Section 2.1.1: probability density decreases monotonically as one moves either higher or lower away from the central value. Therefore, if we can tune one of the arguments to a test predicate we can affect its chance of success. Let DA(q) signify that q is a data approximate quantity. Now let us introduce the notation $PQ_+(p,q)$ to express that the magnitude of the quantity q directly influences the magnitude of the probability of success of test predicate p. Similarly, $PQ_-(p, q)$ indicates that the magnitude of quantity q negatively influences the magnitude of the probability of success of test predicate p. This provides a mechanism for connecting the probability of a predicate being satisfied with the magnitude of a quantity. The rule below is one of several which follow from our definition for data approximations:

$$PQ_-(a < b, a) \Leftarrow DA(q), Q_+(b, q). \tag{4}$$

The rule states: if q is a data approximate quantity and hence uncertain and directly influences the magnitude of a quantity b, the likelihood of $a < b$ succeeding is inversely proportional to the magnitude of a. Let PP(q) indicate that the quantity q is a plan parameter and hence is tunable. Let $PS_\uparrow(p)$ signify that the probability of success of predicate p is increasing. Let $Q_\uparrow(q)$ signify that quantity q is increasing. Therefore we could use the following rules to increase the probability of a predicate p give than q is a plan parameter:

$$PS_\uparrow(p) \Leftarrow PQ_+(p, q), Q_\uparrow(q). \tag{5}$$

$$Q_\uparrow(q) \Leftarrow PP(q). \tag{6}$$

The second rule above asserts that we can increase plan parameters to achieve goals (because they are tunable).

It must also be possible to propagate the qualitative probabilities of predicates. Let $ANT(p_1, p_2)$ indicate that p_2 is an antecedent of a rule for which p_1 is a consequent. One sound rule for propagation of qualitative probabilities across rules can then be expressed:

$$PQ_+(p_1, q) \Leftarrow ANT(p_1, p_2), PQ_+(p_2, q), \forall x[[ANT(p_1, x) \wedge x \neq p_2] \Rightarrow \neg PQ_-(x, q)] \tag{7}$$

The general rules required to construct a qualitative tuning explanation fall into four categories:

general qualitative inference rules – inference rules necessary to reason about increasing and decreasing quantities

Example:

$$Q_\uparrow(x) \Leftarrow Q_+(x, y), Q_\uparrow(y). \tag{8}$$

qualitative predicate definitions – rules providing qualitative definitions for system predicates relating quantities

Example:

$$Q_+(x, y) \Leftarrow [x = y + z]. \left(because(x = f(y)) \wedge \left(\frac{\partial f}{\partial y} = 1 > 0\right)\right) \tag{9}$$

approximation definition rules – rules defining the behavior of test predicates using data-approximate quantities

Example:

$$PQ_{+}(a < b, b) \Leftarrow DA(q), Q_{+}(a, q). \tag{10}$$

qualitative probability rules – rules about the propagation of qualitative probabilities

Example: Rule 7 above.

The qualitative tuning explanation is a sound proof of how to positively influence the probability of success of the predicate which supported the failing expectations. The procedure for constructing the tuning explanation and tuning the plan parameters as a result is as follows:

1. Compute the set P of generalized preconditions and effects for the plan justification structure using the EGGS or EBG algorithms.
2. Take all generalized variables which are quantitative arguments to every predicate $p \in P$ as quantity variables for the qualitative reasoning process.
3. Find all qualitative influences among these quantity variables. This is possible since, if two plan quantities are related, we know the exact functional relationship.
4. Construct a proof based on the qualitative rules discussed above for how to unambiguously qualitatively increase the probability of success of the predicate supporting the expectations to the failed action.
5. Collect the set of quantity increases and decreases justified by the fact that plan parameters PP(q) are tunable. This amounts to finding applications of the rules: $Q_{\uparrow}(q) \Leftarrow PP(q)$ and $Q_{\downarrow}(q) \Leftarrow PP(q)$.
6. For each of the tunable plan parameters in the set collected above, add a new preference to the set of preferences for that parameter. This preference will be incomplete, specifying as one of it bounds a general expression for the point at which the failure occurred and as its relation increasing or decreasing as given in step 5. The other bound will be realized in conjunction with the neighboring preferences.

Next, we give an illustration of step 6 for adding preferences and updating the associated quality function. The overall refinement procedure is demonstrated in more detail with regard to the GRASPER implementation in Section 3.

2.5.2. *Tuning Plan Parameters: An Example*

The actual tuning of a plan parameter q_i is based on the addition of new preferences to the set $P(q_i, C)$. Consider the situation depicted in Figure 5. The goal is to grasp Block2, a relatively flat intricately shaped piece. The workspace also contains Block1, a bigger

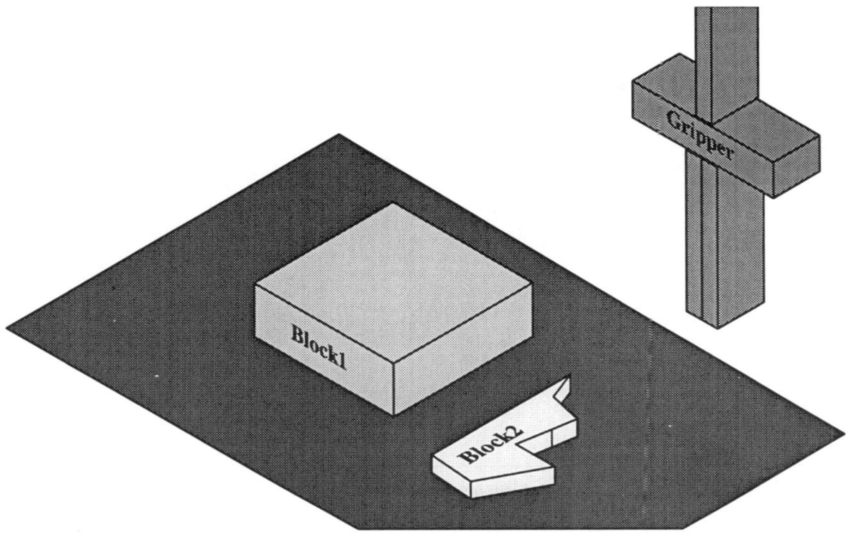

Figure 5. The Gripper In Its Initial Configuration

Current Preferences	Preferences Realized As	Quality Function
<?,?,*constant*>	$a = L_{low}(D(q_{ow}))$, $b = L_{high}(D(q_{ow}))$ <a,b,*constant*>	

Figure 6. Initial Preference Set and Quality Function for the Opening Width Plan Parameter

block which is fairly close to Block2. The only grasping faces on Block2 which can lead to a successful grasp require that one of the gripper fingers be inserted between Block1 and Block2. The system's plan for accomplishing the grasping goal involves selecting a grasping site, moving to a position above the center of the grasping site, orienting to the correct angle for the grasp, opening the gripper to surround the object, moving the gripper down to the table to surround the object, closing the gripper on the object, and lifting the object. Although many parameters are involved in this plan, our example focuses on the parameter for determining proper opening width. The initial preference for the opening-width parameter is shown in Figure 6. In our example we will assume that the initial preference is always for a constant function over the range of the bounds on the parameter. This makes sense in that initially no preferences would imply that all settings (within the bounds) of values for the parameter are equally good. Once the first preference is learned by the system, the initial default preference is removed. The flat quality function defined by the default preference has a continuous range of potential maxima which exist between the

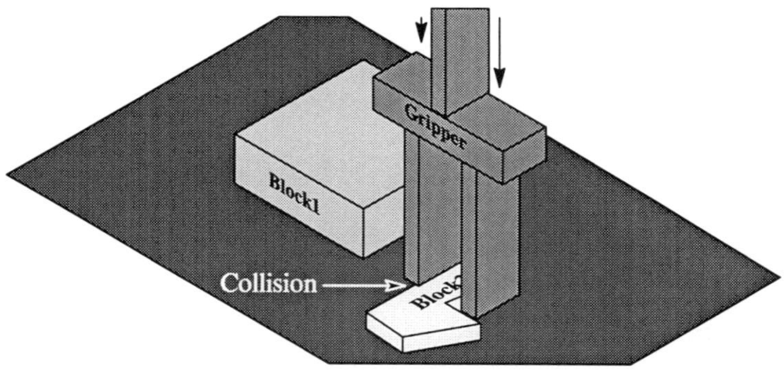

Figure 7. The Gripper Collides With Block2 As It Moves Down To Surround It

Current Preferences	Preferences Realized As	Quality Function
<a,?,*increasing*>	$a = L_{low}(D(q_{ow})), b = L_{high}(D(q_{ow}))$ <a,b,*increasing*>	

Figure 8. Preferences and Quality Function for Opening Width After Collision Failure with Block2

bounds. In our example, the system will select one closest to the current opening width of the gripper. One could have been picked randomly as well but this method has the additional benefit of resulting in less gripper movement. The gripper was initially in a closed position so the nearest setting for the opening-width parameter is one exactly equal to the width of Block2 at the grasping site. The bound a in Figure 6 corresponds with this width.

The plan is executed and fails with an unexpected collision when moving down to surround Block2. Figure 7 shows the situation at the time of failure. The collision occurred near Block2 and at a point lower than the height of Block1. The qualitative tuning explanation therefore indicates that increasing the opening width for surrounding Block2 causes the fingers to be farther from Block2 at that point in the grasp operation and decreases the likelihood of a collision failure with Block2. The collision occurred at point a, a bound on the quality function shown in Figure 6. The tuning explanation suggests that at that point, increasing the parameter is desirable. A new preference is therefore added to the opening-width parameter. Figure 8 shows the new preference and resulting quality function. The algorithm for realizing preferences now assigns the increasing preference to extend over the complete interval because no additional preferences are present. The plan is now more permissive in that the move-down-to-surround portion of the plan should function in spite of errors associated with the data approximations to Block2. However, the system has only learned one of the preferences that are important in this situation. The plan is now applied

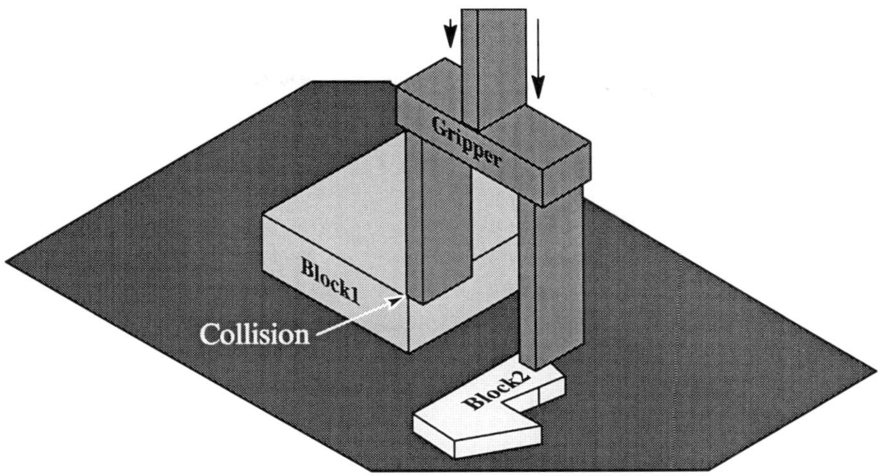

Figure 9. The Gripper Collides With Block1 As It Moves Down To Surround Block2

Current Preferences	Preferences Realized As	Quality Function
<a,?,*increasing*> <?,b,*decreasing*>	$a = L_{low}(D(q_{ow})), b = L_{high}(D(q_{ow}))$ c=(a+b)/2 <a,c,*increasing*> <c,b,*decreasing*>	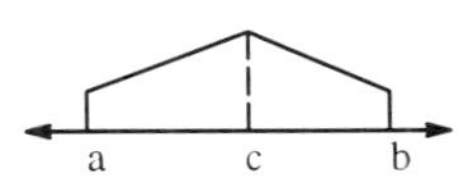

Figure 10. The Preferences and Quality Function for the Opening Width Parameter After a Collision with Block1

again to grasping Block2. The value for the opening-width parameter is now chosen to be the maximum possible opening width prior to striking Block1. This is the upper bound b for the opening-width parameter which corresponds to the only potential maximum of the new quality function shown in Figure 8.

The plan is executed and fails once again. This time the failure occurs near Block1 and at the height of Block1 as shown in Figure 9. The qualitative tuning explanation indicates that decreasing the value of the opening width would decrease the chance of a collision with Block1. As a result, another preference is added indicating that lesser values are to be preferred than b, the point at which the failure occurred. The new set of preferences form a new quality function shown in Figure 10. In order to concretely realize the two incomplete preferences, a third general point c must be created between the bounds a and b. Here, the algorithm creates a general expression for c by taking the midpoint between the values returned by the expressions for a and b, the nearest set of bounds. As a result, the system now has increased the plan permissiveness with respect to errors in the representations for both Block1 and Block2. The failures encountered in this example were with the specific part

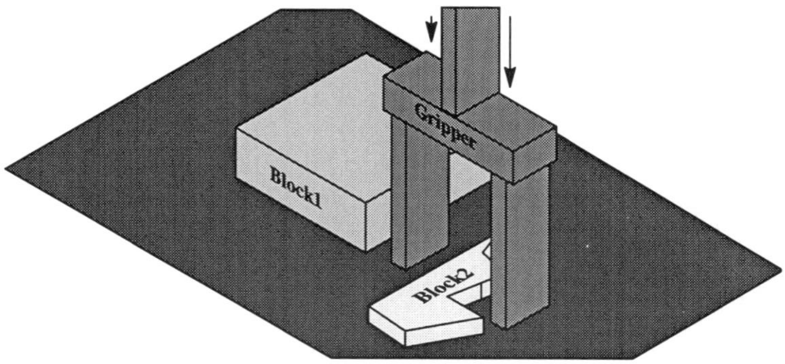

Figure 11. A Successful Surround Operation is Achieved in the Plan

of the plan that moves down to surround the object. It is the permissiveness of this portion of the plan that was increased by tuning the opening-width parameter. Other aspects of the plan which lead to problems will be tuned as well by other encountered failures. The point c now forms the only potential maximum of the new quality function for the opening-width parameter. Consequently, in applying the new plan, a midpoint between the two extremes is chosen for the opening-width parameter. This leads to the successful surround operation in the plan illustrated in Figure 11.

3. The GRASPER System

In order to demonstrate and test our approach we chose a complex real-world domain where uncertainty plays a role: robotic grasping. The goal in this domain is to learn plans which control a robotic manipulator to successfully grasp objects in its workspace. In fact, planning of grasps for arbitrarily shaped objects is still an open problem in robotics. Uncertainty is one primary difficulty in this domain. Sensors don't return precise information. Visual sensors seeking to identify or help represent the objects are very sensitive to placement of the light source. For example, an observer blocking some of the light may actually be effecting visual sensing of the objects. Force sensors used on the manipulator also are subject to errors so that the precise position which the manipulator first contacts an object is not known exactly. The robotic manipulator also cannot be completely precise in its movement. Intractability also plays a significant role in this domain. The system must represent the world in order to construct plans for carrying out its actions. For example, in using visual sensors to model an object or in recognizing objects and retrieving a pre-stored model, that model exists at some resolution. The greater the resolution, the more information that must be considered in planning to grasp the object. Therefore, in order to allow plans to be constructed in a reasonable amount of time, object models must be simplified. As discussed in the previous section, this amounts to introducing some error in return for planning efficiency. Altogether, the robotic grasping domain provides a challenging testbed for learning techniques.

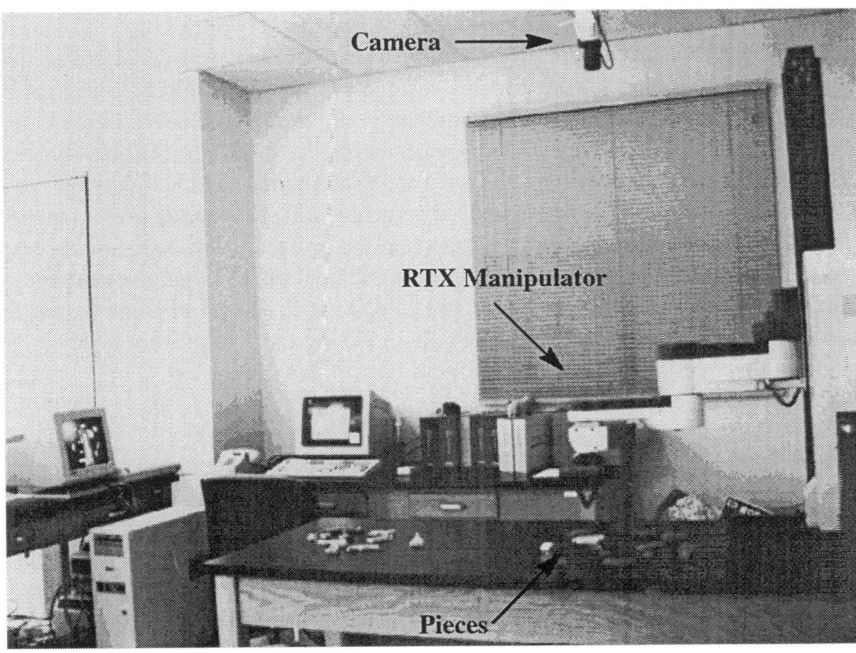

Figure 12. GRASPER Experimental Setup

Figure 12 shows the laboratory setup. The current implementation of the architecture is called GRASPER and is written in Common Lisp running on an IBM RT125. GRASPER is interfaced with a frame grabber connected to a camera mounted over the workspace. The camera produces bitmaps from which object contours are extracted by the system. The system also controls an RTX scara-type robotic manipulator. The RTX has encoders on all of its joint motors and the capability to control many parameters of the motor controllers including motor current. This gives the system a rudimentary capability of detecting collisions with the RTX gripper. If enough current (force) is applied to the motor to overcome friction of the joint and the position encoder indicates no movement, an obstacle has been encountered. This type of sensing gives feedback during execution of a plan when the camera's view of the workspace would otherwise be obscured. This precise control of the manipulator is ideal for carrying out monitored actions in the world.

Our current goal for the GRASPER system in the robotics grasping domain is to successfully grasp the plastic pieces from puzzles designed for young children. Since the pieces are relatively flat and of fairly uniform thickness, an overhead camera is used to sense piece contours. The GRASPER system could be applied to arbitrary three-dimensional objects as well given sufficient sensing hardware to recognize or model the objects in 3D. These pieces have interesting shapes and are large enough, yet challenging, to grasp. The goal is to demonstrate improving performance at the grasping task over time in response to failures. Some of the failures the current implementation learns to overcome, when using isolated grasp targets, include learning to open wider to avoid stubbing the fingers on objects, and learning to prefer more parallel grasping faces to prevent unstable grasps. Detailed examples of these two failures follow later in this section.

3.1. System Architecture

The system is organized as illustrated in Figure 13. There are three modes of operation. Input to the system is through the planning/explanation-based learning component shown in Figure 14. In the simplest mode, a goal is presented to the system which already corresponds to a plan in the knowledge-base whose preconditions are satisfied in the current state of the world. In this case, that plan is passed directly to the executive to be carried out. Secondly, a goal could be presented to the system which doesn't correspond to any known plans. In this case, an explanation is generated for how the goal can be achieved in the current state, the explanation is then generalized and packaged into a general plan which is then saved and passed on for execution. Lastly, a goal and observed action sequence can be given to the planning/ EBL component which then generates the explanation from both the goal and the observed actions. This is the preferred mode of operation of explanation-based learning systems because it can make explanation construction an easier task. In this last mode, the explanation is then generalized, packaged into a plan, saved, and the plan is passed on to the executive.

The executive instantiates the execution sequence associated with the plan with the appropriate bindings obtained from evaluation of the plan's preconditions in the current state. These actions will be monitored, having sensor expectations associated with them that define success or failure. Should a failure occur, as defined by the expectations, the plan

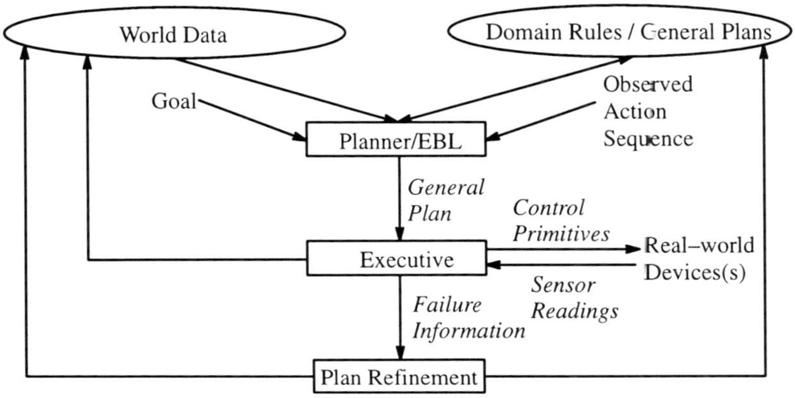

Figure 13. Approximation Architecture

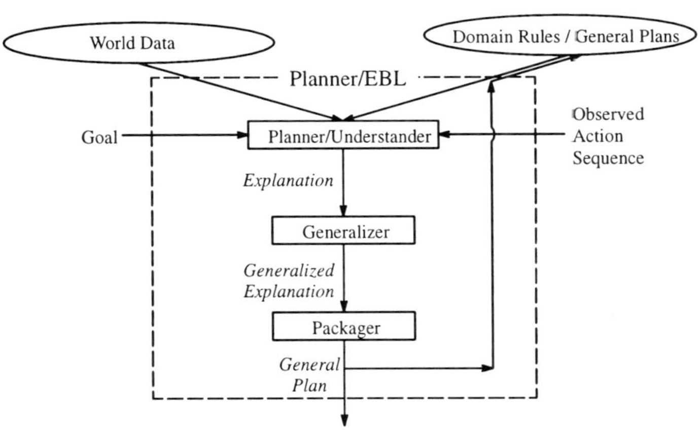

Figure 14. Approximate Explanation-based Learning

```
(MONITOR   (MOVE-GRIPPER ?GRIPPER32878 CLOSE 20 64 20 POSITION)   <- Close Gripper Until Termination Met
           (AND  (FINAL-FORCE GRIPPER ?GFFORCE32731)
                 (> ?GFFORCE32731 50)                           <- Final force should exceed 50 units and final
                 (FINAL-POSITION GRIPPER ?GFPOS32732)              position should be such that the gripper
                 (> ?GFPOS32732 1))                                didn't close on itself (otherwise failure).
           (OR    (AND  (POSITION GRIPPER ?GPOS32733)
                        (PRINT (LIST (QUOTE POSITION) ?GPOS32733))
                        (QUAL= ?GPOS32733 0))                   <- Terminate action if the gripper closes on
                 (AND   (FORCE GRIPPER ?GFORCE132734)              itself or if the force exceeds 60 units
                        (PRINT (LIST (QUOTE FORCE) ?GFORCE132734))
                        (> ?GFORCE132734 60)))
           (STABLE-GRASP ?GRIPPER32878 ?OBJECT32870 ((RELATIVE-FACE ?NAME132744 ?X132762
?Y132763 ?REF-ANGLE32779 ?LEN132748) (RELATIVE-FACE ?NAME232749 ?X232764 ?Y232765 ?REF-
ANGLE32824 ?LEN232753))))   <- Justification for the expectations is that a stable
                               grasp has been planned by the system.
```

Figure 15. An Example of a Monitored Action

refinement module is called to increase plan permissiveness appropriately. If no errors occur, the system is ready for the next goal or goal/observation pair.

3.2. Execution & Monitoring for Robotics

In order for the system to improve its approximations when they don't perform well in the world, the system must have a monitoring capability. It is important that the system be able to represent what actions are to be carried out, what the expected outcome of those actions is, why that outcome is expected, and when the specified actions should be terminated. In the robotic grasping domain, the set of actions to be monitored are a set of motor commands to the manipulator. These may occur as individual motor moves as with a command to move the arm up the column in the case of our SCARA-type manipulator. A group of simultaneously applied motor commands may also be monitored. For instance, in moving the manipulator while grasping an object, force has to be continually applied to squeeze the object while the other joint moves are being carried out. In this case, expectations may apply both to whether the object is sensed between the fingers and whether an external force is sensed by the arm during the motion. Figure 15 gives a concrete example of a monitor employed for closing the robotic gripper on an object. The single action specified is for the gripper to begin closing from its current position. The expectation is that the final force of the gripper on the object exceed 50 units and that the gripper not close on itself. The action terminates when the gripper exerts a force greater than 60 units on the object or the gripper closes on itself. The expectation of feeling the object between the fingers with force greater than 50 units is justified by an explanation for why the planned grasp is stable (so the object will not slip away as force is applied). The specification of expectations as well as their justification facilitates the parameter tuning process.

The expectations and terminations are specified in DNF form and may reference predicates known to the system as well as sensors available on the manipulator. For actual execution on the robot, the sequence of monitored actions specified by a plan is compiled into a Common Lisp program for rapidly checking the sensors while the actions proceed. Many tradeoffs exist in the monitoring process. For instance, since sensors take time to read, the

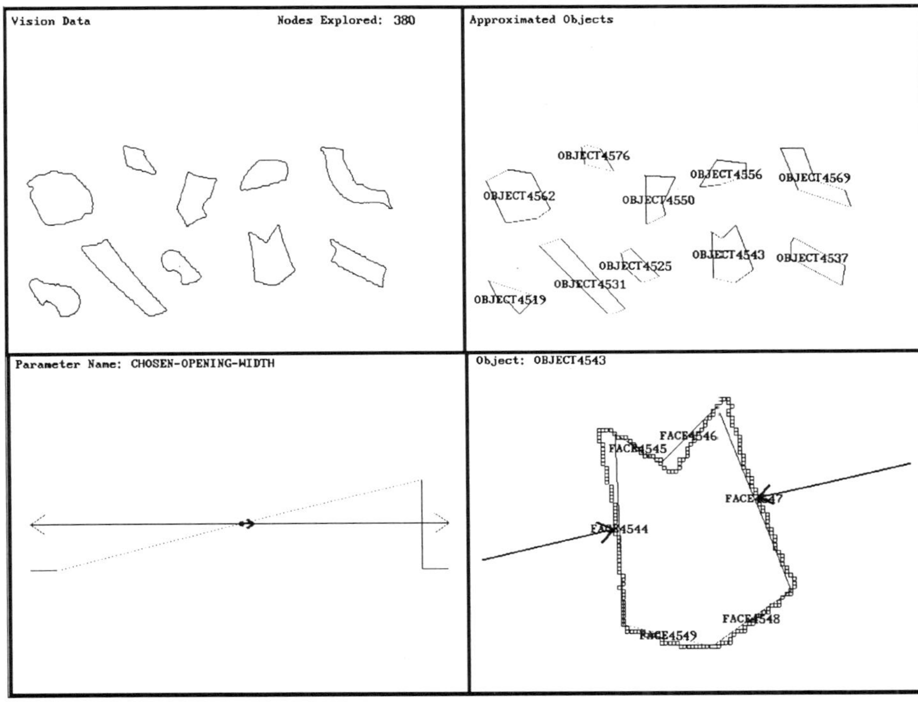

Figure 16. System Status Display During Grasp of Object4543

faster the actions are carried out, the smaller is the number of sensors readings which can be obtained. Furthermore, one can read more types of sensors during execution but each will have a lesser number of readings because of the time constraint. However, because of the permissive planning approach the plans become less sensitive to sensor error.

3.3. Detailed Examples of Learning Episodes

This section details two learning episodes with GRASPER. The first highlights diagnosis of a failure to surround the piece and the second a failure to choose the best grasping site.

3.3.1. Example 1

Figure 16 shows the system's status display during a grasping task. First, the system uses the camera to acquire contour information about objects in the workspace. These contours are shown in the upper left corner of the figure. Next, the contours are approximated with n-gons (internal data approximations) which result in (n2-n)/2 possible unique grasping face

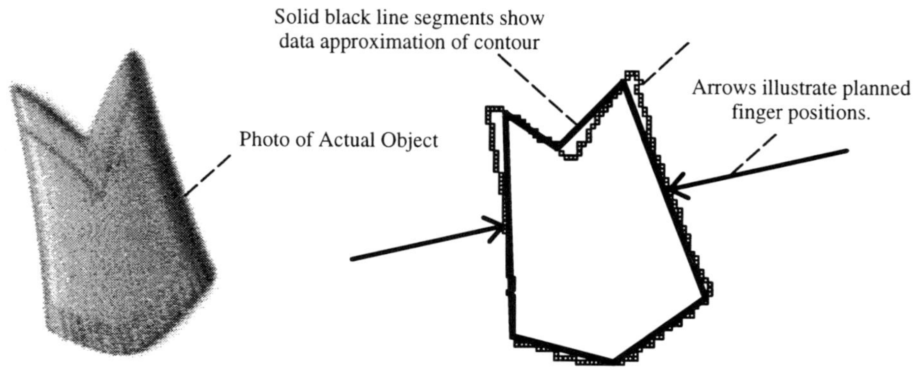

Figure 17. Grasp Target and Planned Finger Positions

pairs. These approximated object contours appear in the upper right quadrant of Figure 16. The algorithm chooses the value of n such that an approximation to the object is possible within a certain error threshold. The data approximated object representations as well as the current information about the state of the robot manipulator are asserted in the initial situation. The target object is then selected and an explanation is generated for how to achieve a grasp of the target. Figure 17 highlights the selected target object. The heavy line indicates the data approximation to the object contour while the lighter pixels show the actual sensed object contour points. The arrows indicate the positions of the leading edges of the fingers for the grasp position given by the produced explanation. The explanation for achieving *grasp-object* involves a total of about 300 nodes with a maximum depth of 10 levels.

Parameters are implemented by rules which choose their value dependent on context. When new preferences are added for a parameter, the associated rules are updated so as to choose potential maxima of the new quality function. Generalized plans refer to the consequents of the rules which choose parameter values. In this case, initially there were no preferences for the plan's opening width parameter other than it be a legal value between the width of the target object and the minimum of the maximum gripper opening and the distance to the nearest object. One of the initial rules for the opening width parameter is shown in Figure 18. This rule pertains to the case where the gripper is currently open less than the minimum opening to satisfy the surround goal in the plan. It then chooses the minimum opening which satisfies the goal because this is the closest potential maxima of the (initially flat) quality function. The rule therefore affects the separation of the arrows shown in Figure 17. After the explanation was generated, and its associated operator sequence executed, the monitored action shown in Figure 19 encounters a violation of the expected sensor readings. Figures 20 and 21 show traces of force and position, respectively, plotted against time (in seconds) into the overall operator sequence.

The original explanation for the *no-gripper-collision-object* goal indicated in the above monitored action is now suspect due to the violated expectations. A sketch of the specific

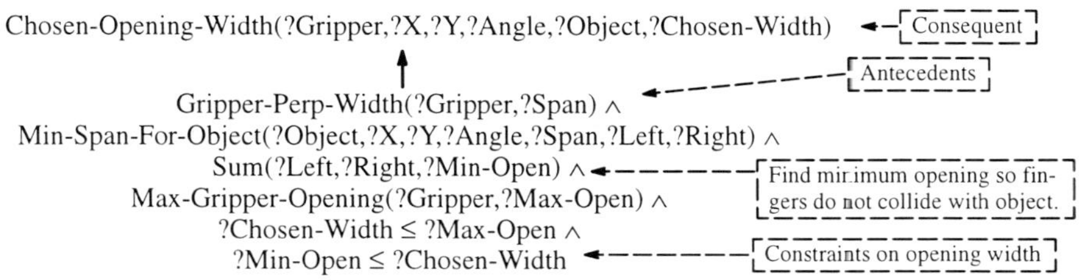

Figure 18. One of the Initial Parameter Rules For Opening Width

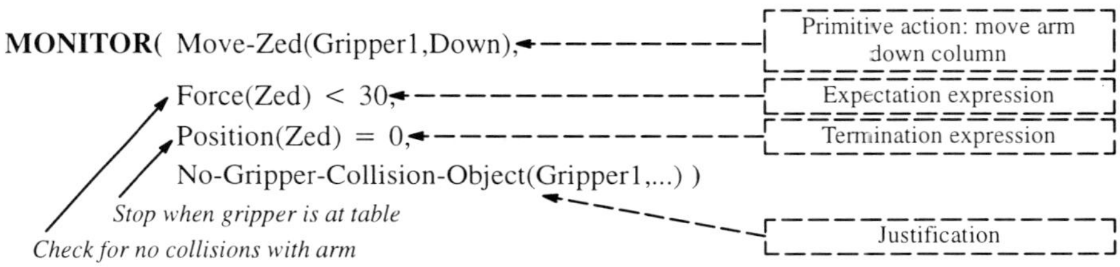

Figure 19. The Failing Monitored Action

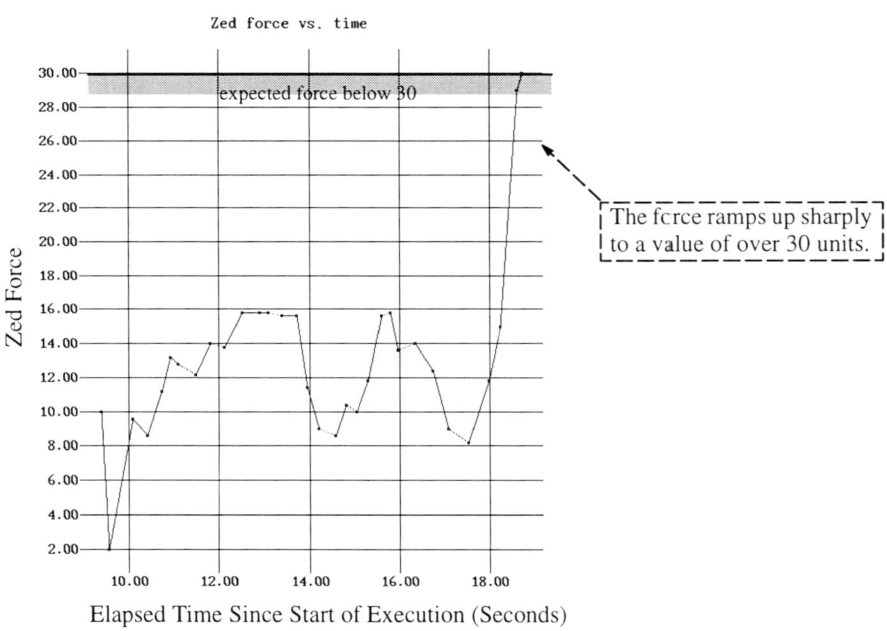

Figure 20. Zed Force vs. Time into Action Sequence

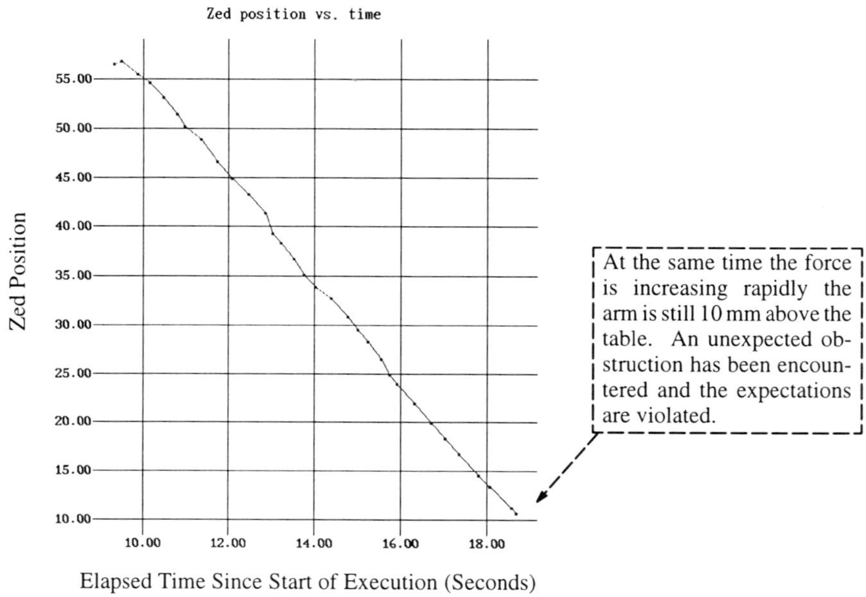

Figure 21. Zed Position vs. Time into Action Sequence

(NO–GRIPPER–COLLISION–OBJECT GRIPPER1 289.62 267.53 –12.70 38.89 OBJECT4543)
- (LEFT–FINGER–OF GRIPPER1 FINGER1)
- (NON–INTERSECTING–GRIPPER–FINGER–OBJECT GRIPPER1 FINGER1 289.62 267.53 –12.70 38.89 OBJECT4543)
  - *Subproof for translating finger to appropriate opening width (6 facts, 8 built–ins)*
  - *Subproof for counter–rotating object center for clipping against finger (8 built–ins)*
  - *Subproof for calculating extents and checking for overlap (7 built–ins)*
- (RIGHT–FINGER–OF GRIPPER1 FINGER2)
- (NON–INTERSECTING–GRIPPER–FINGER–OBJECT GRIPPER1 FINGER2 289.62 267.53 –12.70 38.89 OBJECT4543)
  - *Subproof for translating finger to appropriate opening width (6 facts, 8 built–ins)*
  - *SHARED Subproof for counter–rotating object center for clipping against finger (8 built–ins)*
  - *Subproof for calculating extents and checking for overlap (7 built–ins)*

Figure 22. Explanation Specific to Failure

explanation is shown in Figure 22. This explanation for why no external force should have been sensed during the downward move of the gripper is the starting point for developing the qualitative tuning explanation. Data approximate quantities and tunable parameters employed in the plan support proof (explanation) are identified and asserted as such. A proof

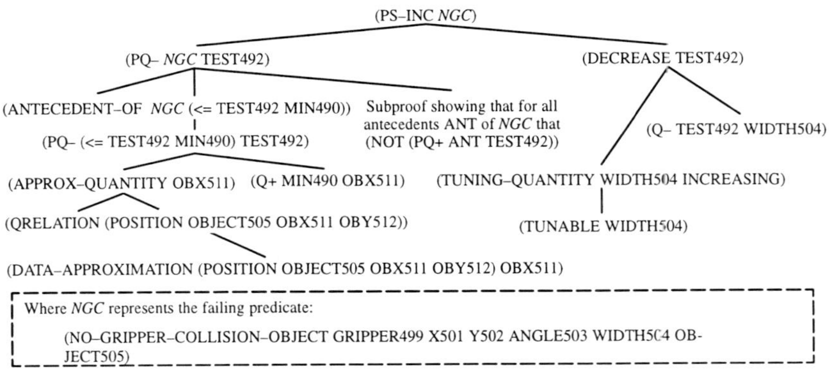

Figure 23. A Qualitative Tuning Explanation

is then constructed for increasing the probability of success of the *no-gripper-collision-object* goal. Figure 23 shows the qualitative explanation for how opening the gripper (increasing the opening-width tunable parameter) positively influences the probability that there will be no collision between the first gripper finger and the object. Table 1 gives the semantics for the predicates employed in the explanation. The topmost left-hand subtree establishes that decreasing the quantity TEST492 can positively influence the probability of success of the the probability of the *no-gripper-collision* predicate. This is because decreasing the quantity TEST492 can increase the probability of the predicate ($\leq$ TEST492 MIN490) which is an antecedent of a rule supporting the *no-gripper-collision* predicate and all of the other antecedents to that rule can be shown as non-decreasing with respect to decreasing TEST492.[4] The probability of success of the predicate ($\leq$ TEST492 MIN490) increases when TEST492 is decreased because MIN490 is influenced by a data approximate quantity. The right subtree of the proof establishes that the quantity TEST492 can be decreased because it is influenced inversely by a tunable parameter WIDTH504 which can be increased. The parameter WIDTH504 is the opening width parameter for the gripper.

The qualitative tuning explanation indicates that the chosen-opening-width parameter should be tuned. An increasing preference will be posted at the minimum opening width chosen in the failure. Figure 24 illustrates the shape of the chosen-opening-width parameter's quality function before (left) and after (right) tuning has occurred. After parameter tuning, the rules associated with parameter are updated Afterwards, the rule associated with this parameter reads a shown in Figure 25. This rule prefers selection of the peak of the newly re-calculated quality function which corresponds to opening as wide as the current situation permits. When the new more permissive plan is applied the resulting gripper finger positions are as illustrated in Figure 26.

(PS–INC ?pred)	*the magnitude of the probability of success of ?pred is increasing*
(ANTECEDENT–OF ?pred1 ?pred2)	*?pred2 is an antecedent of ?pred1 in the rule being analyzed*
(PQ– ?pred ?quant)	*the magnitude of the quantity ?quant negatively influences the magnitude of the probability of success of predicate ?pred*
(PQ+ ?pred ?quant)	*the magnitude of the quantity ?quant positively influences the magnitude of the probability of success of predicate ?pred*
(DECREASE ?quant)	*the magnitude of the quantity ?quant is decreasing*
(APPROX–QUANTITY ?quant)	*?qant is a data approximate quantity*
(Q+ ?q1 ?q2)	*the magnitude of quantity ?q2 positively influences the magnitude of quantity ?q1*
(Q– ?q1 ?q2)	*the magnitude of quantity ?q2 negatively influences the magnitude of quantity ?q1*
(TUNING–QUANTITY ?a ?b)	*plan parameter ?a is tuned according to the relation ?b*
(QRELATION ?pred)	*?pred is one of the predicates employed in the plan support structure (and employs continuous quantities which can be modelled qualitatively)*
(DATA–APPROXIMATION ?pred ?arg)	*the argument ?arg of the predicate ?pred is known to be a data approximate quantity*
(TUNABLE ?quant)	*the magnitude of quantity ?quant is a tunable plan parameter*

Table 1. Predicates Employed in the Tuning Explanation

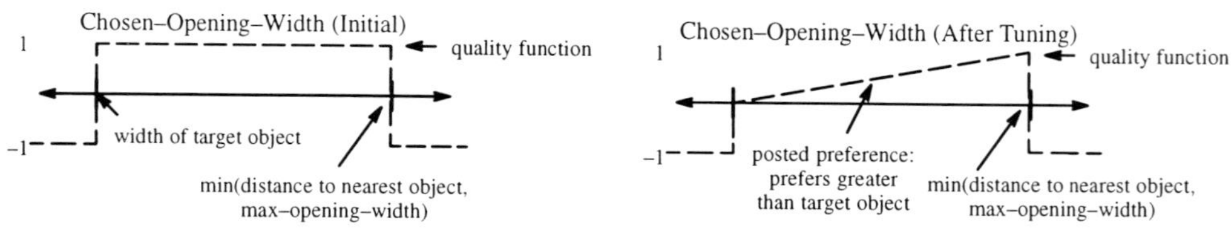

Figure 24. The Chosen-Opening-Width Parameter Quality Function After Learning New Preference

```
INTRA-RULE: R4611
   FORM:
      (CHOSEN-OPENING-WIDTH ?GRIPPER ?X ?Y ?ANGLE ?OBJECT ?RETURN)
   ANTS:
      (GRIPPER-PERP-WIDTH ?GRIPPER ?SPAN)
      (DISTANCE-TO-CLOSEST-OBJECT ?OBJECT ?X ?Y ?ANGLE ?SPAN ?RADIUS)
      (GRIPPER-FINGER-PARALLEL-WIDTH ?GRIPPER ?PSPAN)
      (DIF ?RADIUS ?PSPAN ?NRADIUS)
      (MAX-GRIPPER-OPENING ?GRIPPER ?MAX-OPEN)
      (MIN ?NRADIUS ?MAX-OPEN ?RETURN) <— peak chosen here
      (MIN-SPAN-FOR-OBJECT ?OBJECT ?X ?Y ?ANGLE ?SPAN ?LEFT ?RIGHT)
      (SUM ?LEFT ?RIGHT ?MIN)
      (<= ?MIN ?RETURN) <— must be wider than object
   CONS:
   PARAMETER: CHOSEN-OPENING-WIDTH
```

Figure 25. Rule Supporting Opening-Width Parameter After Tuning

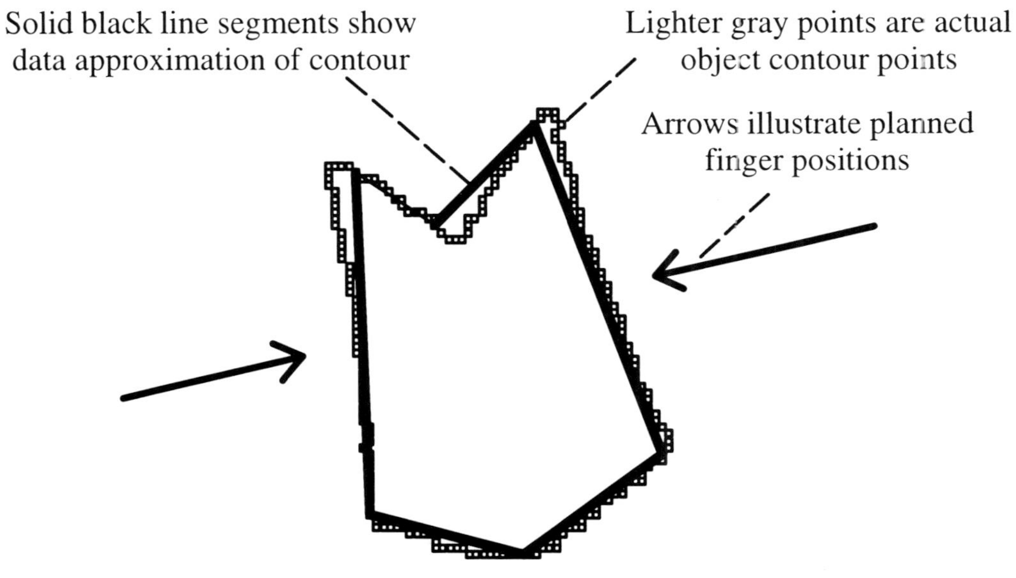

Figure 26. A Successful Wide Grasp

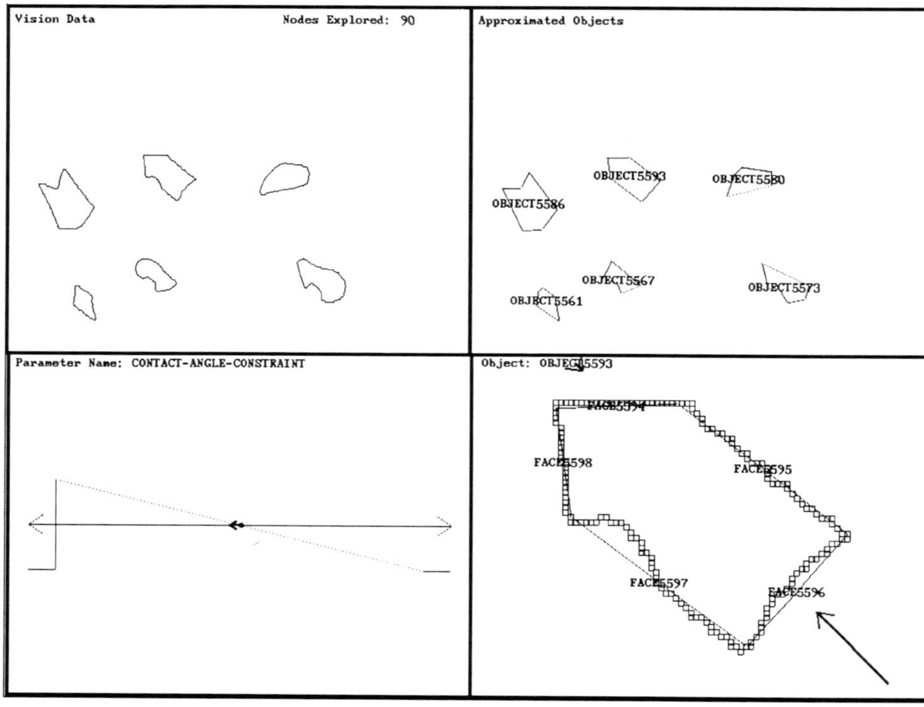

Figure 27. System Status Display During Grasp of Object5593

3.3.2. Example 2

Next, after the system has learned to open widely when grasping objects, a second object is presented for grasping. Figure 27 shows the system status display while initially planning the grasp for this object. Figure 28 highlights the selected target object. The dark line indicates the polygonal object approximation and the light colored pixels show the sensed object contour points. The arrows illustrate the planned positions for the fingers in the generated grasping plan. Notice, that the fingers are well clear of the object due to the tuned opening-width parameter learned in the first example. Generally, the opening-width parameter is the first one to be tuned because striking the objects while attempting to surround them is a fairly common error. However, the parameter regarding acceptable angles between contact faces still has only the initial default preference. The angle between chosen faces must be greater than 0 (parallel) and less than the angle at which slipping occurs according to the friction coefficient between the gripper and faces being grasped. All angles fitting these criteria are treated as equally good. Consequently, the two faces chosen for this grasp should be acceptable given the specified friction coefficient. The rule on which the contact-angle-constraint parameter is based is:

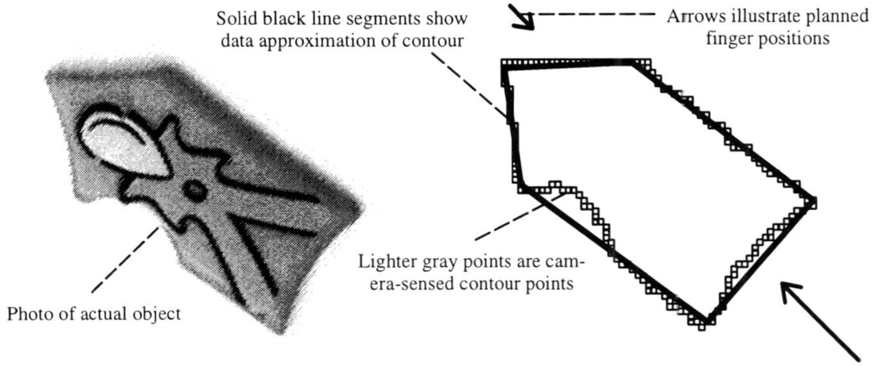

Figure 28. Grasp Target and Planned Finger Positions

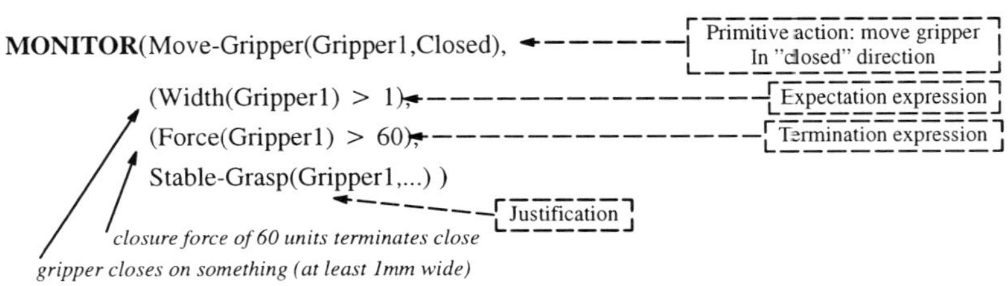

Figure 29. The Failing Monitored Action

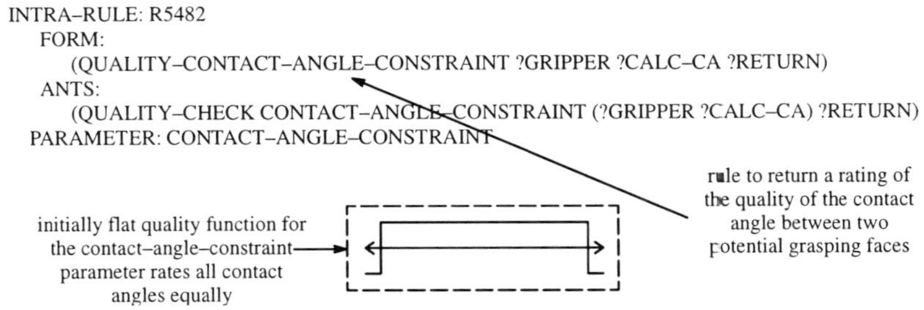

This rule references a built-in predicate QUALITY-CHECK which returns a rating of the specified contact angle based on the current quality function for that parameter. The rating is used in choosing from among several candidate contact angles and is therefore only good in relation to other contact angles with respect to the same quality function. After the explanation is generated, and its associated operator sequence executed, the monitored action shown in Figure 29 encounters a violation of the expected sensor readings. The traces in Figure 30 and Figure 31 show, respectively, plots of gripper force and gripper position with time (in seconds) into the overall execution sequence.

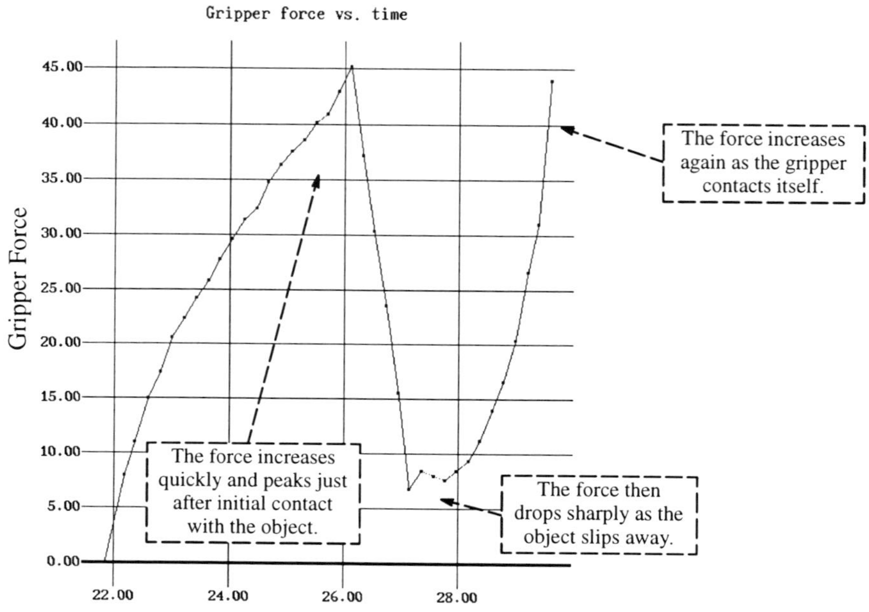

Figure 30. Gripper Force vs. Time into Action Sequence

The original explanation for the *stable-grasp* goal indicated in the above monitored action is now suspect due to the violated expectations. A sketch of the specific explanation is shown in Figure 32. This explanation for why a stable grasp should have been achieved is the starting point for developing the qualitative tuning explanation. Data-approximate quantities and tunable parameters employed in the plan support proof are identified and asserted as such. A proof is then constructed for increasing the probability of success of the *stable-grasp* goal. Figure 33 shows the qualitative explanation for how preferring a smaller contact angle positively influences the probability that a stable grasp will be achieved. Table 1 gives the semantics for the predicates employed in the explanation. The topmost left-hand subtree establishes that as the quantity A747 is decreased the probability of success of a stable grasp is increased. This is true because the predicate ($\leq$ A747 FANGLE635) is an antecedent of the rule supporting the stable grasp, the probability of success of ($\leq$ A747 FANGLE635) is influenced negatively by the magnitude of A747, and none of the other antecedents to the rule are influenced positively by the same quantity. The probability of success of ($\leq$ A747 FANGLE635) is influenced negatively by A747 because FANGLE636 is influenced by a data-approximate quantity FRIC634, the friction coefficient between the object faces and gripper. The topmost right-hand subtree shows that the quantity A747 is a tunable parameter which can be decreased.

The qualitative tuning explanation indicates that smaller contact angles should be preferred in choosing grasping faces. Figure 34 illustrates the shape of the contact-angle-constraint

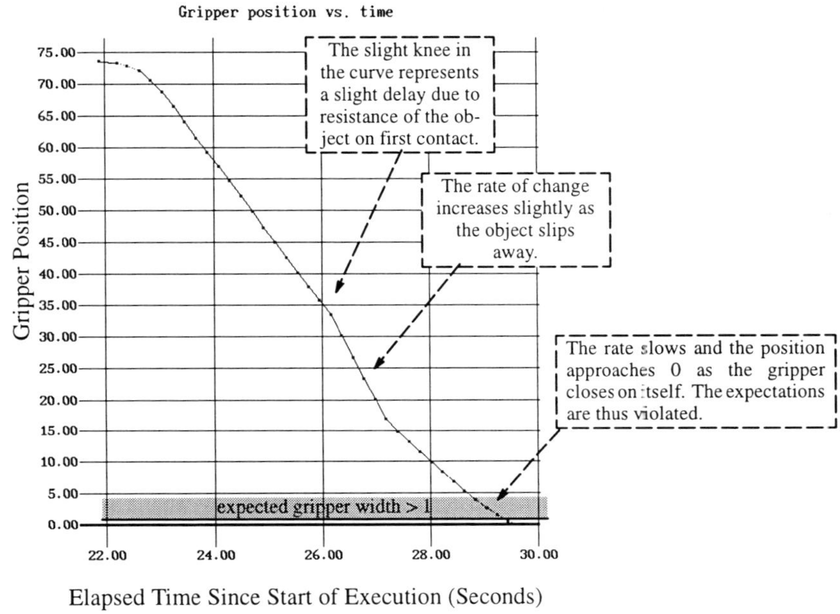

Figure 31. Gripper Position (Width) vs. Time into Action Sequence

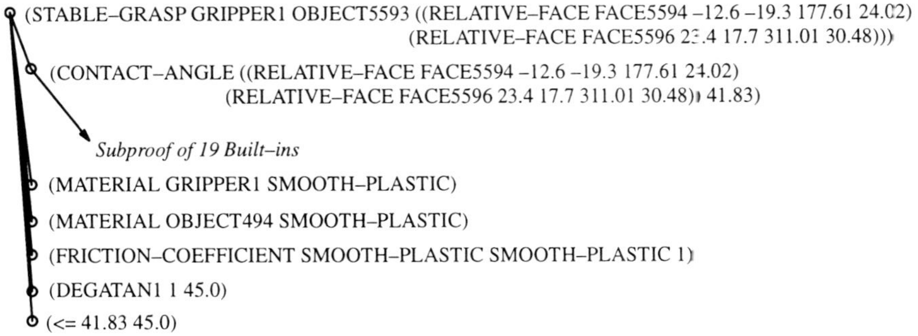

Figure 32. Explanation Specific to Failure

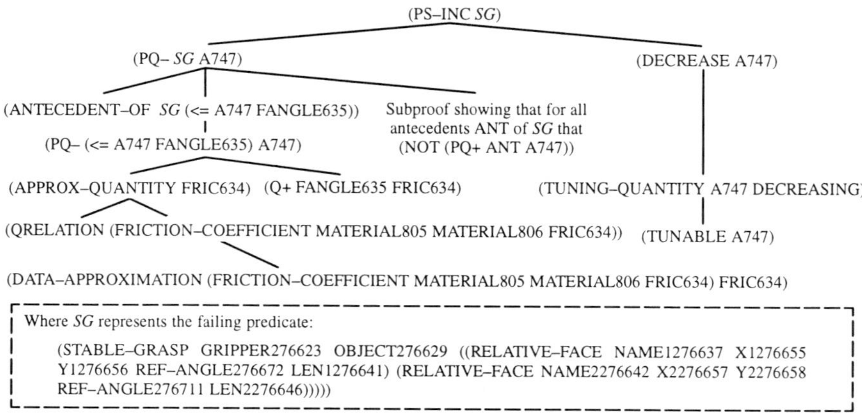

Figure 33. A Qualitative Tuning Explanation

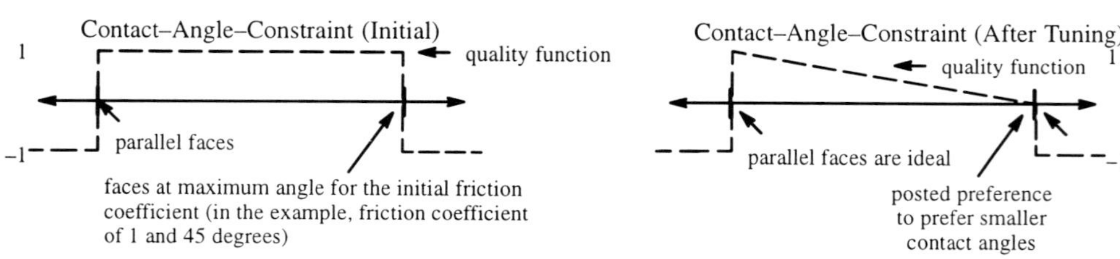

Figure 34. The Contact-Angle-Constraint Parameter Before and After Tuning

parameter's quality function before (left) and after (right) tuning has occurred. After parameter tuning, the rule which uses the parameter to rate contact faces (shown earlier) will take advantage of the new contact-angle-constraint parameter's quality function and choose a grasp position such that the two faces to be contacted are closest to parallel. Figure 35 shows the new more permissive plan as applied to the same object. In this case, the most parallel faces were preferred over those picked in the earlier application of the plan.

4. Empirical Results

The GRASPER system was given the task of achieving equilibrium grasps on the 12 smooth plastic pieces of a children's puzzle. Figure 36 shows the gripper and several of the pieces employed in these experiments. A random ordering and set of orientations was selected for presentation of the pieces. Target pieces were also placed in isolation from other objects. That is, the workspace never had pieces near enough to the grasp target to impinge on the decision made for grasping the target. The first run was performed with parameter

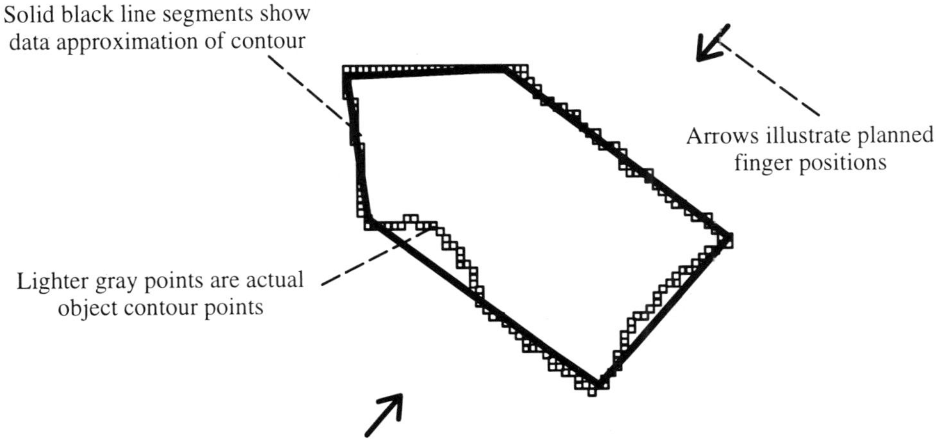

Figure 35. Successful Grasp Employing Tuned Parameter Constraining Contact Angle

Figure 36. Gripper and Pieces

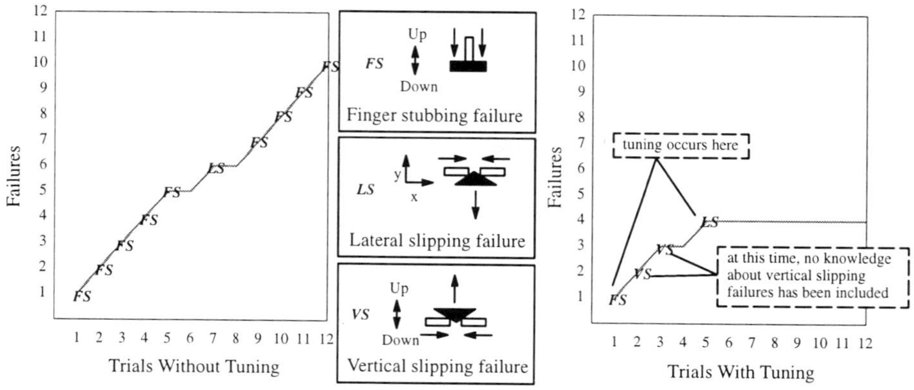

Figure 37. Comparison of Tuning to Non-tuning in Grasping the Pieces of a Puzzle

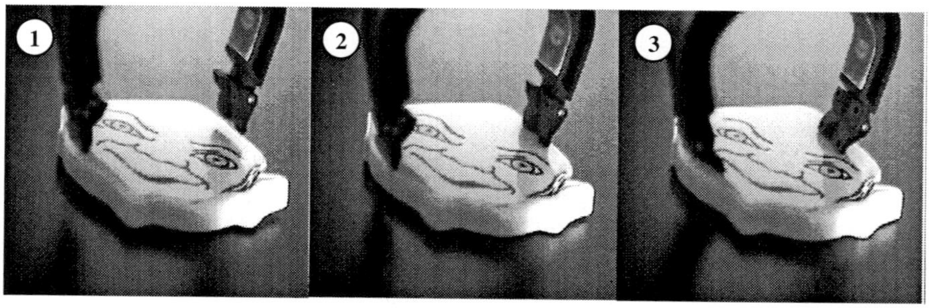

Figure 38. An Instance of a Finger Stubbing Failure

tuning turned off. The results are illustrated in Figure 37. Failures observed during this run included *finger stubbing failures* (see Figure 38) where a gripper finger struck the top of the object while moving down to surround it and *lateral slipping failures* (see Figure 39) where, as the grippers were closed, the object slipped out of grasp, sliding along the table surface. In the system's initial approximate representation for the world, the choice of grasping faces is constrained only by the gripper being able to open wide enough to surround them and that an equilibrium grasp is realizable with the current gripper-object friction coefficient (initially 1 here). Since a friction coefficient of 1 is likely to be high for these materials, the choice of contact faces is likely to be under-constrained initially, resulting in slipping failures. The choice of opening width is the minimum deviation from the current opening width (initially 0 here) which satisfies the approximate model of the grasp. Due to uncertainties in the world, this approximate opening width may often result in stubbing failures. Therefore, the error rate of our initial approximate plan was high resulting in 9 finger stubbing failures and 1 lateral slipping failure in 12 trials.

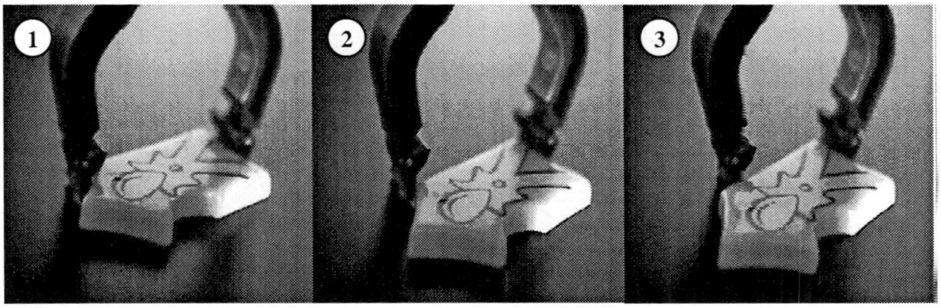

Figure 39. An Instance of a Horizontal Slipping Failure

In our second run, parameter tuning was turned on. An initial stubbing failure on trial 1 led to a tuning of the chosen-opening-width parameter which determines how far to open for the selected grasping faces. Since the generated qualitative tuning explanation illustrated that opening wider would decrease the chance of this type of a failure, the system tuned the parameter to choose the largest opening width possible (constrained only by the maximum gripper opening and possible collisions with nearby objects). In the case of isolated grasp targets, opening to the maximum gripper width is preferred. In trials 2 and 3, finger stubbing failures didn't occur as they had previously because the opening width was greater than the object width for that orientation. Vertical slipping failures, which the current implementation does not currently have knowledge about, did occur. The system had to be told that a vertical slipping failure had occurred instead of the lateral slipping failure it thought had occurred. This is because, without further knowledge about vertical slipping failures and a means for detecting them, they look in other ways (the force vs. position profile of the gripper closing) like a lateral slipping failure. Preventing vertical slipping failures involves knowing shape information along the height dimension of the object, which we plan to give in the future using a model-based vision approach. In trial 5, a lateral slipping failure is seen and the qualitative tuning explanation suggests decreasing the contact angle between selected grasping surface through tuning the contact-angle-constraint parameter. This is tuned to prefer smaller contact angles. A single tuning for the finger stubbing and lateral slipping failures was sufficient to eliminate those failures with isolated grasp targets.

5. Conclusions & Future Work

In the work discussed here, an assumption is made about limited interactions between plane parameters. We plan to eliminate this assumption by extending the method to learn about interactions between parameters. The illustration shows a view from above of two objects in the robot's workspace. The black rectangles are the gripper fingers as seen from above. The goal is to grasp object 1. The fingers are positioned along the pair of faces chosen for this grasp. Factors like the angle between the fingers and faces and friction coefficient will be the same no matter where the gripper contacts the object along that particular pair of faces.

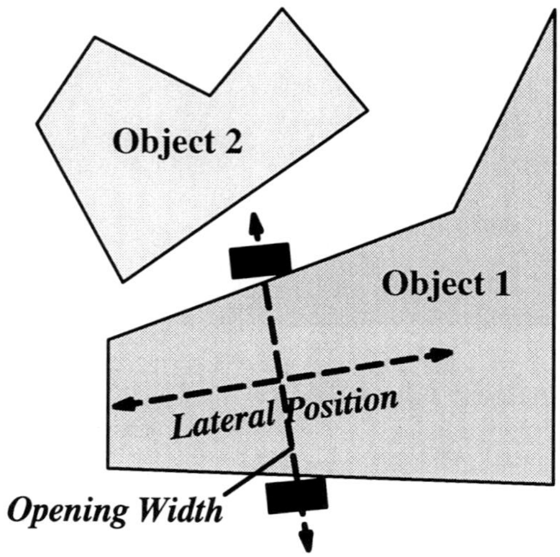

Figure 40. An Example of a Parameter Interaction: Lateral Position and Opening Width in the Robotic Grasping Domain

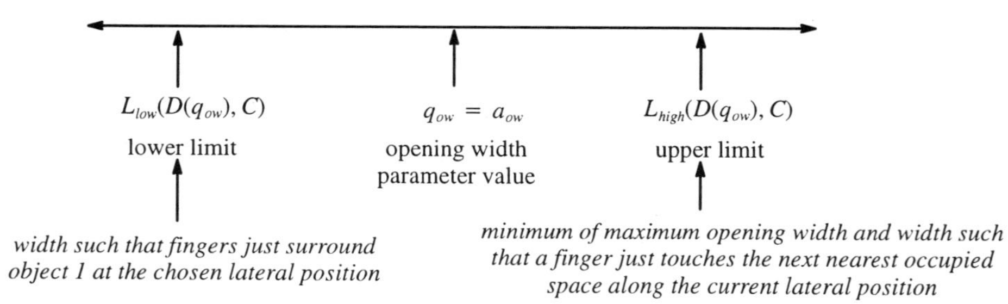

Figure 41. A Graphic Representation of Limits on Opening Width

However, the lateral position along the pair of faces affects the choice of opening width both because of the changing width of object 1 as one moves laterally and because of the nearby object 2. The upper and lower bounds on the opening width parameter depend on the lateral position parameter. These are illustrated in Figure 41. Suppose that the plan constructed by the system chooses a value for the lateral position parameter prior to choosing a value for opening width. Under the limited interaction assumption made currently this assumes that a lateral position is chosen such that $L_{low}(D(q_{ow}), C) \leq L_{high}(D(q_{ow}), C)$ and therefore a possible opening width is guaranteed. For a planner to guarantee this assumption requires exhaustively reasoning about all interacting parameters, an expensive process. It makes sense to learn incrementally about interactions which lead to failures. Instead of spending the time to guarantee no interactions, our initial plan could choose a lateral position without regard to opening width. Now suppose a lateral position was chosen such that the width of object 1 at that point was wider than the maximum opening width of the gripper. In order to prevent trouble with the interaction, the constraint on opening width would be imposed on the choice of lateral position. The savings of an incremental approach to management of interactions is based on the assumption that many interactions, although theoretically possible are not likely to arise in practice. For instance, suppose all the objects in the workspace tend to have a maximum dimension well within the maximum opening width of the gripper and are well separated from nearby objects. In this case, the lateral position vs. opening width interaction is not likely to be a factor.

Another interesting area for study is the possibility of relaxing the conditions for generation of a qualitative tuning explanation. Guaranteeing soundness of such explanations can be costly. The requirements for generating qualitative proportionalities (Q_+ and Q_-) could be relaxed to permit transitivity of influences. For example, from $Q_+(a,b)$ and $Q_+(b,c)$ we could conclude $Q_+(a,c)$. Although such rules can make qualitative proportionalities easier to derive, they can introduce ambiguities. That is, it may be possible to show both $Q_+(a,b)$ and $Q_-(a,b)$. Qualitative tuning explanations (QTEs) would then be more heuristic than guaranteed methods of demonstrating a way to increase the probability of success of the failed predicates. However, such heuristic QTEs could be produced very quickly. The value of these heuristic QTEs would be tested empirically. Further heuristic QTEs could be produced if initial ones fail in practice. This approach is very much in the flavor of plausible EBL (DeJong, 1989).

The current approach assumes that every failure is worth preventing. This is not necessarily true. If we had reason to believe that a failure encountered was very unlikely, it may not be in our interest to attempt to tune parameters to prevent it. First, tuning parameters does require some effort which may outweigh the possible gain in tuning the unlikely failure. Second, tuning parameters to prevent the failure may involve some cost in terms of the modified plan. For instance, the parameter tuning may incur extra motion of a manipulator which results in slower execution of the overall plan. One possible strategy for deciding what failures are worth tuning is to empirically perform initial testing in the domain to discover what failures occur most frequently. Another approach is to develop a weak theory of errors which can be used to justify which failures are worth tuning.

When a newly learned preference is inconsistent with the existing preferences for some plan parameter, it becomes necessary to distinguish the context in which the conflicting preference occurred from the current context. We call this *context splitting*. A new set of preferences can then be created with the new preference added and the conflicting preference removed which is applicable to the current context (as distinguished from the context attached to the original set). One method of distinguishing contexts is to ask the user (Rosenbloom, Laird & Newell, 1987). We would like to have the system make the distinction using its domain knowledge. The knowledge is available all along but, without experience in the world, the system has insufficient experience to conclude which distinctions are important. In our framework, inconsistent preferences trigger the process of context splitting.

Notes

1. While experiments have not been performed with incorrect qualitative theories it seems that the technique would work in spite of some types of erroneous qualitative theories while it would result in poor performance with others.
2. For a model of the different aspects of utility for a plan to be executed in uncertain, complex domains see (Bennett, 1989).
3. Q_+, Q_-, and the associated inference rules about increasing and decreasing quantities are used as in Qualitative Process Theory (Forbus, 1984).
4. The subproof supporting the expectations is packaged into a single general rule prior to generation of the qualitative tuning explanation.

References

Agre, P. & Chapman, D. (1987). Pengi: An implementation of a theory of activity. In *Proceedings of the National Conference on Artificial Intelligence*, pages 268–272.

Bennett, S.W. (1989). Learning uncertainty tolerant plans through approximation in complex domains. Master's thesis, University of Illinois at Urbana-Champaign.

Brooks, R.A. (1982). Symbolic error analysis and robot planning. Technical Report 685, MIT AI Lab.

Brooks, R.A. (1987). Planning is just a way of avoiding figuring out what to do next. Technical Report 303, MIT AI Lab.

Chapman, D. (1987). Planning for conjunctive goals. *Artificial Intelligence*, 32(3):333–378.

Davis, E. (1986). *Representing and Acquiring Geographic Knowledge*. Morgan Kaufmann Publishers, Inc.

DeJong, G.F. (1989). Explanation-based learning with plausible inferencing. In *Proceedings of The Fourth European Working Session on Learning*, pages 1–10.

DeJong, G.F. & Mooney, R.J. (1986). Explanation-based learning: An alternative view. *Machine Learning*, 1(2):145–176.

Erdmann, M. (1986). Using backprojections for fine motion planning with uncertainty. *International Journal of Robotics Research*, 5(1):19–45.

Firby, R.J. (1987).. An investigation into reactive planning in complex domains. In *Proceedings of the National Conference on Artificial Intelligence*, pages 202–206.

Forbus, K.D. (1984). Qualitative process theory. *Artificial Intelligence*, 24:85–168.

Hammond, K., Converse, T. & Marks, M. (1990). Towards a theory of agency. In *Proceedings of the Workshop on Innovative Approaches to Planning Scheduling and Control*, pages 354–365.

Hutchinson, S.A. & Kak, A.C. (1990). Spar: A planner that satisfies operational and geometric goals in uncertain environments. *Artificial Intelligence*, 11(1):30–61.

Hwang, Y.K. (1988). *Robot Path Planning using Potential Field Representations*. PhD thesis, University of Illinois at Urbana-Champaign.

Khatib, O. (1986). Real-time obstacle avoidance for manipulators and mobile robots. *International Journal of Robotics Research*, 5(1):90–98.

Lozano-Perez, T., Mason, M.T. & Taylor, R.H. (1984). Automatic synthesis of fine-motion strategies for robots. *International Journal of Robotics Research*, 3(1):3–24.

Malkin, P.K. & Addanki, S. (1990). Lognets: A hybrid graph spatial representation for robot navigation. In *Proceedings of the Eighth National Conference on Artificial Intelligence*, pages 1045–1050.

Mitchell, T.M., Keller, R. & Kedar-Cabelli, S. (1986). Explanation-based generalization: A unifying view. *Machine Learning*, 1(1):47–80, January 1986.

Mitchell, T.M., Mahadevan, S. & Steinberg, L.I. (1985). Leap: A learning apprentice for vlsi design. In *Proceedings of the Ninth International Joint Conference on Artificial Intelligence*, pages 573–580.

Mooney, R.J. & Bennett, S.W. (1986). A domain independent explanation-based generalizer. In *Proceedings of the National Conference on Artificial Intelligence*, pages 551–555.

Rosenbloom, P.S., Laird, J.E. & Newell, A. (1987). Knowledge level learning in soar. In *Proceedings of the National Conference on Artificial Intelligence*, pages 499–504.

Schoppers, M.J. (1987). Universal plans for reactive robots in unpredictable environments. In *Proceedings of the Tenth International Joint Conference on Artificial Intelligence*, pages 1039–1046.

Segre, A.M. (1988). *Machine Learning of Robot Assembly Plans*. Kluwer Academic Publishers.

Suchman, L.A. (1987). *Plans and Situated Actions*. Cambridge University Press.

Wong, E.K. & Fu, K.S. (1985). A hierarchical orthoganal space approach to collision-free path planning. In *Proceedings of the 1985 Institute of Electrical and Electronics Engineers International Conference on Robotics and Automation*, pages 506–511.

Zadeh, L.A. (1965). Fuzzy sets. *Information and Control*, 8(3):338–353.

Zhu, D.J. & Latombe, J.C. (1990). Constraint reformulation in a hierarchical path planner. In *Proceedings of the 1990 Institute of Electrical and Electronics Engineers International Conference of Robotics and Automation*, pages 1918–1923.

Machine Learning, 23, 163–189 (1996)

Robot Programming by Demonstration (RPD): Supporting the Induction by Human Interaction

H. FRIEDRICH, S. MÜNCH AND R. DILLMANN {friedric|smuench|dillmann}@ira.uka.de
University of Karlsruhe, Institute for Real-Time Computer Systems and Robotics, Kaiserstr. 12, D-76128 Karlsruhe, Germany

S. BOCIONEK AND M. SASSIN {bocionek|sassin}@zfe.siemens.de
Siemens AG, Corporate R&D, ZFE ST SN 6, Otto-Hahn-Ring 6, D-81739 Munich, Germany

Editors: Judy A. Franklin, Tom M. Mitchell, and Sebastian Thrun

Abstract. Programming by Demonstration (PbD) is a programming method that allows software developers to add new functionalities to a system by simply *showing* them in the form of few examples. In the robotics domain it has the potential to reduce the amount of time required for programming and also to make programming more "natural". Just imagine the task of assembling a torch by a manipulator. Wouldn't it be nice to just assemble the torch with one's own hands, watched by video and laser cameras and maybe wearing data gloves, i. e. sensors that provide the data to automatically generate the necessary robot program for the assembly task? And wouldn't it be even nicer to demonstrate the task with few different torches, but achieving an assembly function for all possible variants of them?
In order to realize such a PbD environment, at least two major problems have to be solved. First, the sensor data have to be transformed to high-level situation-action descriptors, a task that is not yet solved in general. Second, if a generalization is required, induction algorithms must be applied to the recorded and transformed traces, aiming to find the most general user-intended function from only few examples. In this article we will concentrate only on the second problem. The described experimental environment consists of an industrial robot (PUMA 260b), a 6D teach mouse input device, and some sensors. Various data can be recorded during a demonstration for further processing in the PbD system running on a workstation. The objective is to explore the possibilities of integrating learning and clustering algorithms for automated robot programming. In particular it is investigated how human interaction within system- as well as user-initiated dialogs can support the induction component.

Keywords: Programming by Demonstration (PbD), Robot Programming by Demonstration (RPD), Teaching, Robots, Manipulators, User Intentions, Dialog-Based Learning (DBL)

1. Introduction

Programming by Demonstration (PbD) is an approach for faster software development in domains where the user works interactively with a computer or some machinery. Such domains are the programming of graphical user interfaces as well as graphics and text editors where the user interacts with the editor program by using keyboard, mouse, pen, or other input devices. Another important domain is the programming of robots, appliances, and other machines where the user moves and manipulates the device by provided control panels. In all cases the goal of PbD is to "show" the system a new functionality by demonstrating the desired behavior, rather than coding it in a programming language. What has been shown is recorded as traces of sensor data, transformed into basic actions, analyzed, and then a computer program is generated that allows the user to repeat the demonstrated task by calling one command. To make the program even more useful, additional generalization methods have to be applied during the generation of the function.

Imagine the problem of assembling something with a robot manipulator. To program such a task today, often so-called teach-in methods are used. A "programmer" directs the robot's end-effector to all necessary locations, where it has to grip or manipulate something, by means of six buttons on a teach panel (three buttons for x-, y-, and z-direction, three buttons for the respective orientations). At each important position the teacher stores the location as a 6-tuple, and additional commands like open or close gripper. Replaying the whole sequence of stored "button hits" repeats exactly the taught program. Teach-in programming of manipulators can take a lot of time and is really acceptable only if large numbers of exactly the same product items shall be assembled.

Programming by demonstration with some generalization capabilities overcomes the uneconomic usage of teach-in especially for the manufacturing of a product with many possible variants. Let us take as a most advanced scenario the assembly of a torch. Wouldn't it be comfortable to just assemble it with one's own hands, watched by video and laser cameras and maybe wearing data gloves, i. e., sensors that provide the data to automatically generate the necessary robot program for the assembly task? And wouldn't it be even more comfortable to demonstrate the task with few different torches, but achieving an assembly function for all possible variants of them?

Unfortunately, PbD settings as described in this scenario are not yet available neither in industry nor at universities or are only partially realized. The reasons are manifold: Sensors are still too expensive, and the sensor processing of complex image sequences is not fast and accurate enough to fit the strong needs of many manufacturing processes. A human arm and hand with their more than 30 degrees of freedom (DOF) are more flexible in performing object manipulations than an ordinary 6-DOF robot with its gripper. Another problem is that one single human is mostly not sufficient to demonstrate complex production or assembly sequences. There is always a variety of interactions between machines, robots, and humans. In addition, not all steps of an assembly are usually performed at one place. The whole process is a sophisticated sequence of time-sliced actions at different stations along a conveyor belt. Therefore, programming the whole torch assembly task by demonstration can probably not be done as sketched in the scenario.

However, there are industrial "subtasks" that could already be programmed by demonstration today—even with generalization—and would significantly reduce the required programming effort. Such subtasks are pick&place macros for different settings, the loading of pallets, sorting tasks, part feeding tasks, etc. They are far from complicated assembly tasks, but still need the same time-consuming and awkward teach-in programming procedure. (Instead, one could also use graphical off line programming, but that method also needs a significant amount of teach-in postprocessing in order to adjust the generated program to the real setting (Neubauer et al., 1992).)

Robot subtasks like pick&place or sorting objects often contain the need for the generalization of geometrical positions, of constants to variables, and of similar objects to classes. There is also a need for finding loops or recursive calling schemata in the action traces. Such generalizations are well-known from inductive machine learning, in particular explored under the label of *learning from examples*. In fact, as in most practical applications (see the discussion in (Thrun & Mitchell, 1993)), a combination of analytical and inductive learning is used for robot programming by demonstration (RPD). This means that previous knowledge of the domain and particular knowledge about tasks and environments have to be available.

Nevertheless, PbD and RPD have to cope with inherent problems that do not necessarily exist for all learning-from-examples tasks. There are only *very few* examples available, typically one to five, because nobody wants to demonstrate the same task very often. This causes great uncertainty about what the user intended with the demonstrated examples, particular because the hypothesis spaces are large. As a way to overcome this problem we propose dialog-based learning (DBL) (Bocionek & Sassin, 1993), a means to engage the demonstrating person in the induction process by question&answer dialogs that may be initiated by the RPD system as well as by the person. The usefulness of DBL has already been shown in other PbD domains in which—based on only few examples—functions for graphics editors (Sassin, 1994) and personal assistance software (Bocionek, 1995) could be learned (induced).

The remainder of this article is organized as follows. In section 2 we will describe the principle of robot programming by demonstration in more detail and show our solution for an adequate system architecture. Section 3 contains the description of our RPD setting and describes in detail the two major phases of RPD. In section 4 we will give a detailed example where human interaction is used during the induction process. A discussion of the work related to our approach follows in section 5. Finally, section 6 concludes the paper and outlines planned future work.

2. Robot programming by demonstration

> *The key characteristic of robots is versatility; (...). (...), but it can be exploited only if the robot can be programmed easily. In some cases, the lack of adequate programming tools can make some tasks impossible to perform. In other cases, the cost of programming may be a significant fraction of the total cost of an application. For these reasons, robot programming systems play a crucial role in robot development.*
>
> (Tomás Lozano-Pérez, 1983)

Having in mind Lozano-Pérez' demand for an adequate robot programming environment, we want to give a short overview on common programming methods and describe the principles of our approach, called Robot Programming by Demonstration (RPD).

2.1. Traditional robot programming methods

Today, many different programming methods with a large number of variants exist. In (Lozano-Pérez, 1983), three major categories can be identified:

1. guiding;
2. explicit or robot-level programming;

3. implicit or task-level programming.

In *guiding* (also known as *teaching by showing, teach-in, or play-back*), the operator moves the manipulator and records explicitly all relevant positions. The major advantages of this method are that only positions are taught that can really be reached by the robot and that the explicit recording leads to very precise trajectories. Furthermore, this method is very easy to perform. However, it has some major drawbacks: The robot cannot be used for the production process during the teaching phase and the taught sequence cannot be altered; it is usually even impossible to include sensor signals in the generated program. Additionally, only the robot used for the teaching can exactly follow the recorded path[1].

Explicit or *off line programming* in a programming language introduces control structures like loops and branches, and offers the chance to react to external events by using sensors, thus leading to very flexible robot programs. For the programming process, either special-purpose languages like VAL II (**V**ersatile **A**ssembly **L**anguage) or general-purpose languages with special extensions, like PasRo (**Pas**cal for **Ro**bots), are used. These languages offer much more flexibility than teach-in, and the robot can be programmed off line, i. e. the production process does not have to be interrupted. But, due to the richness and power of these languages, programs are hard to write by robot operators who often perform the teach-in process today. Therefore, usually only highly qualified persons can cope with this method.

The most advanced of the three above-mentioned methods for the programming of robots is *implicit programming*. This method requires a sophisticated planning system, because no paths or positions have to be specified, but the programmer only determines the goals and subgoals of a task which are then further detailed by the planner until the robot's execution level is reached. Problems with this method arise due to the high complexity of the tasks which require planning systems that are not available for industrial usage today.

To complicate the situation, all three methods can be extended and combined. For example, in spot or arc welding the data of the CAD model is used and adjusted with teach-in methods to generate the robot's path. To simplify the programming, graphical simulation systems are used in combination with other methods. This *off line programming* offers the chance to test several alternatives and to visualize the simulated execution. As an extension of graphical simulation, virtual environments (VE) are explored for robot programming, too (see, e. g. (Flaig et al., 1994)).

With robot programming by demonstration (RPD), we want to develop a method that offers richness, power, and flexibility, but is as easy to use as simple teach-in. The principles of RPD are described in the next section.

2.2. Principles of RPD

Robot programming by demonstration is an intuitive method that makes the programming of robots easier and faster for experts, but which can also be used by robot operators due to its simplicity. The idea of RPD is to combine teach-in with learning and to integrate the user in all processing and induction steps.

Learning in an RPD system can be realized at different levels of abstraction. Therefore, skill as well as knowledge acquisition by the system via human demonstration might be integrated in the system (Friedrich & Kaiser, 1995). The different levels are shown in fig. 1.

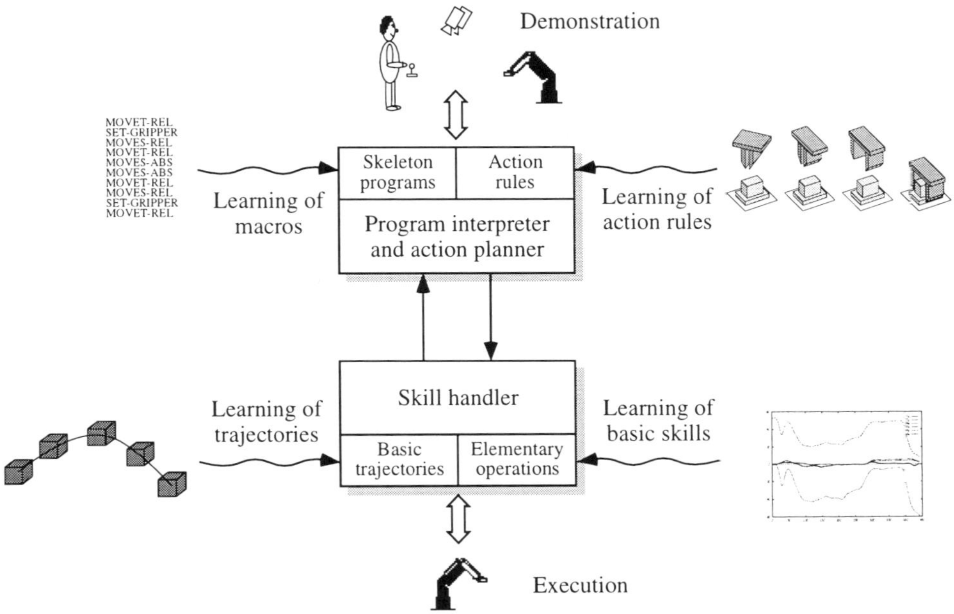

Figure 1. The principle architecture for an integrated RPD system. The learning components are emphasized in the sketch, showing the major input data for the different modules.

The two major levels of an integrated RPD system are as follows:

1. *Learning on the symbolic level:* The system can either generalize robot programs completely or partly, using sequences of elementary operations as input data for the learning process (Münch et al., 1994; 1994b); or the initial acquisition, extension, and refinement of robot action knowledge and object descriptions may be realized by processing data extracted from experiments (Kreuziger & Hauser, 1993).

2. *Learning on the subsymbolic level:* On the lower level, either trajectories can be reconstructed from freehand motions (Ude, 1993; Ude et al., 1994) or robot controllers performing basic skills might be generated with the help of artificial neural networks (Kaiser et al., 1994).

In the system described within this article, we will mainly concentrate on learning robot programs on the symbolic level.

The system architecture developed for the realization of the RPD prototype system is shown in fig. 2. The user interface on the left consists of the 6D mouse for moving the robot, and of some windows on a workstation that enable the user to guide the induction by supervising the generalizations proposed by the system and supplying her or his intentions.

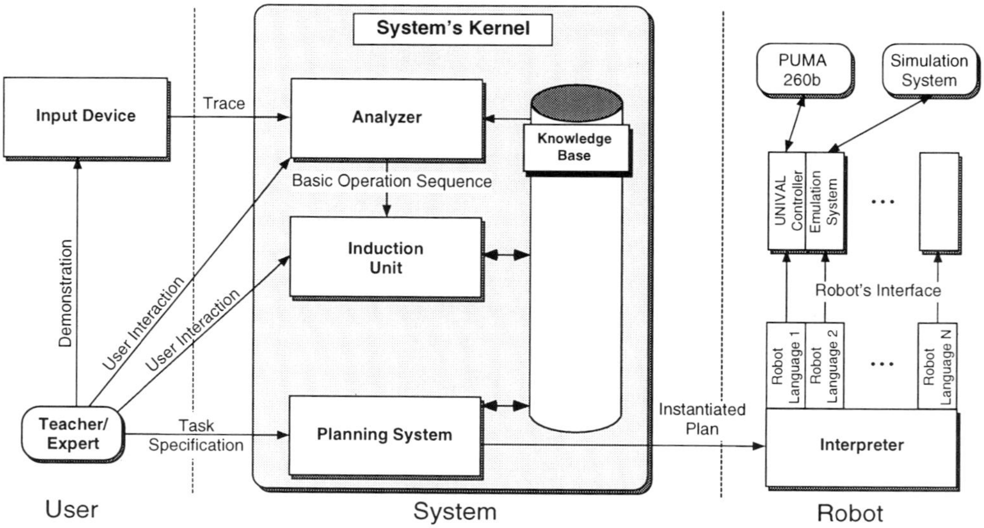

Figure 2. Architecture of the realized RPD prototype system: The interfaces to the user (left) and the robot system (right) are enclosing the RPD kernel system.

The system's kernel is located in the center of the figure. The *analyzer* component performs the required preprocessing steps on the recorded demonstration trace. This preprocessing consists mainly of segmenting the trace and classifying the segments with respect to the given *planning language*. This planning language is basically structured like a conventional programming language, consisting of branches, loops, and *Basic Operations (BOs)*. However, by applying the analyzer only BOs are directly generated from the trace. All control and hierarchical structures are introduced by the *induction component* which operates on the planning language only. The induction of branches and variables is supported by the user. The result of the induction is stored in a knowledge base for later execution as a single plan, but also for use as a planning macro by the execution unit.

The right part of fig. 2 shows the robot's environment for executing the generated program. The program is in a kind of "neutral" format at this point, therefore it has to be transformed into one of several robot programming or simulation languages in order to execute it either with a real robot or with a simulation system.

Three major phases can be identified in the RPD system developed so far: Firstly, the *demonstration phase* in which a solution for the task is demonstrated by the user. All data of this demonstration is stored in a trace for further processing. Thereafter the analyzer component is applied to transform the stored trace on the symbolic level, namely to the representation given by the planning language used in the succeeding components of the system. Secondly, the *induction phase* takes place. Therein the output of the analyzer is taken and processed with respect to additional information—the user's intentions acquired through a man-machine interface. Finally, the *execution phase* is performed in case a previously generated program shall be applied. This includes selecting, sequentializing, and parametrizing the program as well as driving the selected robot. All three phases will be described in more detail in the following section, whereas an example for the whole process is shown in section 4.

3. The RPD environment

After the principles of RPD have been outlined, this section describes the function and the architecture of the RPD system which has been realized prototypically at the Institute for Real-Time Computer Systems and Robotics, University of Karlsruhe (see fig. 3). The techniques and methods underlying the three already mentioned processing phases are described in detail. Preceding this description, a short overview of the planning language used by the system's modules is given.

Figure 3. A sample RPD testbed with a PUMA 260b robot and a 6D bowl, realized at the Institute for Real-Time Computer Systems and Robotics, University of Karlsruhe.

The Planning Language

The language providing the symbolic representation used for generating and processing programs from demonstrations basically consists of five *Basic Operations (BOs)* that are:

`grip(O)`, `ungrip(O)`, `approach(s, d)`, `depart(s, d)`, and `transfer(s, d)`.

Here, O denotes either an object of the world model or a relational condition that can be evaluated and unified with the world model in order to determine an object. s and d are frames (also called *locations*) describing positions and orientations of the manipulators tool center point.

Like most procedural programming languages, the planning language includes higher level elements that represent branches and loops. These are `if-then-else` branches as well as `while` and `repeat` loops. Their body holds a sequence of other operators that can be branches, loops, or even whole plans themselves. A `plan` is an operator containing a sequence of operators and representing the solution to a given task. It can itself be used as part of another plan, branch, or loop. Currently, the loop operators are not used. However, they are already defined for the next development stage of the system.

All operators of the planning language are modeled in the tradition of STRIPS operators (see Fikes & Nilsson, 1971), i. e. for every operator its applicability depends on the evaluation of a *precondition* and after the application a certain *postcondition* holds.

Conditions themselves are expressed in terms of combinations of disjunctions and conjunctions of relations. These are describing relationships between objects like `on(O1, O2)` and `in(O1, O2)` or object-features like `obj-type(O, type)`. Terms are always given and transformed into *disjunctive normal form (DNF)*. They can be evaluated with respect to the world model that is given initially and updated each time one of the operators described above is applied. The implemented variable concept allows to use variables as attributes of relations. Unification of a term holding variable attributes allows to determine all combinations of variable instantiations on the current world model that yield a positive evaluation of the term.

3.1. Demonstration Phase

Demonstrating a task

In order to generate a *new* robot program, the PbD process starts with the *demonstration of an example*. Several different methods exist for that step, each with particular strengths and drawbacks which will not be discussed here in detail. In the system presented here, a 6D mouse is used to control the PUMA 260b robot (see fig. 3), because only a short training period is needed to generate good demonstration data.

Filtering

Pauses occurring in demonstrations and unsteady motions due to not perfectly trained users result in noisy data[2]. Therefore, a *filtering* of the recorded trace follows the demonstration in order to remove redundant vectors to reduce the amount of data (see fig. 7).

Analysis

The next step is the *analysis* of the filtered trace, which is mainly used to transform the "raw data" into the symbolic format of the planning language. To achieve a correct transformation, i. e. the recognition of the correct sequence of BOs hidden in the recorded vector sequence, two problems have to be solved. Firstly the trace has to be segmented into subsequences, each of them representing a certain BO. Secondly each segment has to be classified and mapped on one of the five BOs. Obviously the two tasks of *segmentation* and *classification* depend on each other and therefore have to be done simultaneously. In this sense the transformation task is similar to the problem of continuous speech recognition. Another aspect also common to the speech recognition domain is that of time variations. Just like people speak with different speeds, various users perform demonstration movements with different speeds. Moreover, the duration of a movement usually depends on the distance between the start and the final location.

In the domain of speech recognition *Time Delay Neural Networks (TDNNs)* have been proven to be powerful tools for various segmentation and classification problems (Waibel et al., 1989), overcoming also the problem of time variations in the input signal. For this reason the application of TDNNs was considered for the solution of the analysis problem in the RPD system. For each of the BOs a TDNN was generated, consisting of three layers with 5 neurons in the input and exactly one in the output layer. The number of neurons in the hidden layer varies for the different BOs as well as the applied time-delays and training epochs, as can be seen from table 1.

Table 1. Some parameters of the TDNNs for Basic Operation recognition

BO Net	Perceptrons in the hidden-layer	Time-delays on 1st layer	Number of training epochs
transfer	9	10	10,000
approach	10	15	12,000
depart	10	15	12,000
grip	8	8	10,000
ungrip	8	8	10,000

Motions are restricted to translational movements and each net was trained with nine to ten training sequences (10–20 vectors per training sequence) of pick&place operations. The demonstrations vector sequence is fed simultaneously into each BO TDNN. By studying the outputs produced, i. e. the net activations, a decision about which BO was performed during what time-span of the demonstration is taken.

Automatic correction of classification errors, like a missing gripper event between a detected approach and depart pair is done by applying some heuristics about possible operation sequences in a post-processing step. Following this stage the transformation from the subsymbolic to the symbolic level takes place. The determined sequence of Basic Operations of the planning language is generated and instantiated with respect to the segmented vector sequence.

For manual correction of classification errors and supervision of the analysis module's results, a graphical user interface is provided (see fig. 9). The manipulator's trajectory is plotted divided in its x, y and z coordinates along the time axis. Moreover, the boundaries of the recognized BOs are given. Now the user has the possibility to correct the BO sequence in order to improve the classification results. Collected corrections from multiple demonstrations might be used later to retrain the TDNNs and thus improve the automatic analysis performance of the system over time.

3.2. *Induction Phase*

Constructing a plan from the BO sequence generated by the *analysis* module is a task consisting of the following phases that are all performed by the *induction* unit:

1. Formation of *Object Groups (OGs)*, where each OG contains a subset of the BO sequence that represents the manipulation of exactly one object.
2. Determination of the valid postcondition of each BO sequence of step 1.
3. Acquisition of the *user's intention* regarding the demonstration via man-machine communication.
4. Detection and generation of branches, i. e. conditioned manipulations and elimination of superfluous manipulations.
5. Acquiring the *object selection criteria* for the manipulated objects and introducing variables based on user interaction.

These steps will be discussed in more detail in the following paragraphs.

Preparing BOs; Grouping & Calculating Conditions

For further processing of the given BO sequence, it is structured into subsets, the already mentioned Object Groups[3]. At this stage of the system's development, there is no interaction of the manipulated object with other objects like, e. g., insertion operations. Therefore, division and grouping of the BO sequence into a sequence of OGs is fairly simple. The graph in fig. 4 represents the possible sequences of BOs with and without interactions for a one arm robot. Following this graph OGs are built in a simple straightforward way.

Starting from the `transfer` towards an object, proceeding with its `grip`, the sequence of Basic Operations is traced until the object is finally `ungripped` and the manipulator `departed` from it. Therefore, each Object Group encloses a subpart of the users demonstration that is characterized by the manipulation of one specific object.

As can be seen in fig. 4, the consideration of interactions of objects enables multiple cycles of `transfers` to other objects and interactions of the `gripped` object with those. These interactions take place in between the `grip` and `ungrip` of the primarily manipulated object.

After the identification of all Object Groups, a forward trace over the BO sequence is performed. In this process the fully parametrized pre- and postconditions of every BO are determined. Also the postconditions of every Object Group and finally the whole BO sequence, i. e. the demonstration itself, are determined. Every type of BO has certain preconditions and postconditions that have to be valid for its application respectively are valid after application. In addition, there are pre- and postconditions that depend on the environmental context in which the BO is executed. If, for example, the postcondition for the BO `approach(s,d)` has to be determined, the result depends on whether an object has already been grasped or not. If the gripper does not hold an object (i. e., it is free), `approach`

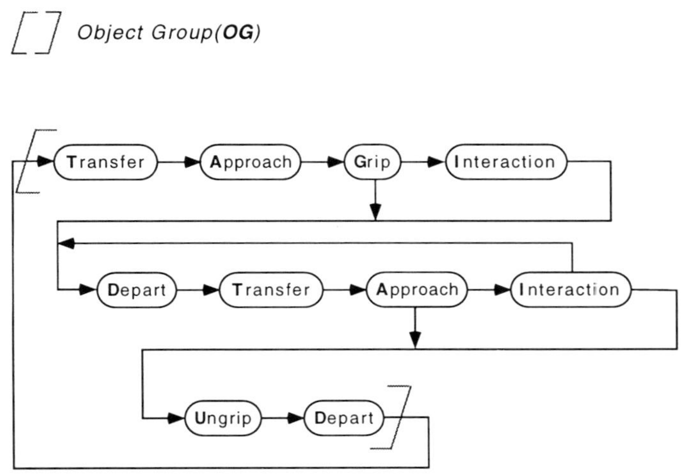

Figure 4. Graph representation of possible BO sequences

is prior to a gripping action. Therefore, the gripper encloses the object `o` to be gripped at the end of the `approach`. This forces the insertion of the `enclosed(O,Manipulator)` relation to the postcondition of the `approach`. A special case is given when an object is ungripped. Here, all possible relations to all other objects have to be tested and added to the `ungrip` BO's postcondition.

While forward tracing through the BO sequence and determining the conditions of the Basic Operators, the world model is updated by virtually applying the BOs to it. Starting with the first BO and the initial environmental configuration, the world model data is updated sequentially after the determination of each operators conditions with respect to the operators definition.

As soon as the conditions of all BOs related to an OG are determined, they can be fused to the postcondition of the group. Naturally this includes to check for relations built up by one BO that where destroyed later by another BO related to the same Object Group. The same fusing of postconditions is done for all OGs in order to determine the whole demonstration's postcondition. In other words, the demonstration's postcondition is the accumulation of all lasting changes that occured throughout the demonstration.

But do all these changes really reflect exactly the users intention? Or were some actions simply forced by the environmental conditions present at that moment? Or were some of the manipulations even unnecessary in order to solve the given task?

Acquiring new Knowledge by User Interaction

Unfortunately a computer system is not able to *guess* the answers to the questions raised above. Since the primary focus in the proposed PbD system is not to induce generalizations and/or knowledge from a large number of examples, a different direction than classical ML induction techniques has to be followed to generate the desired answers. In order to keep the number of required demonstrations as small as possible another source of information has to be integrated in the induction process—the user itself.

The postconditions of the demonstration that are related to the same Object Group are presented to the user successively. Thus he can select the conditions that reflect his intention best and can discard the others. Thereby the system gathers the information necessary to identify actions which are conditioned with respect to the user's intention. With this knowledge, branches can be derived and `if-then-else` operators can be generated. Parts of the demonstration that where found to be superfluous can even be deleted.

Generalizing and Optimizing; Branching & Garbage Collection

While the processing strategy in the first phase of the induction process was characterized by forward tracing through the input BO sequence, the sequence is now traced in the reverse direction, i. e. backwards. Tracing backward, branching, determination of preconditions of the Object Groups, and optimization by deletion of superfluous manipulations is done.

In the following, the developed branching algorithm is given in pseudo-code in which the initializations and update rules are given precisely. *I* denotes the users intention, *UI* unsatisfied intentions, *UPrec* unsatisfied preconditions, *PostC*(og_i) the postcondition and *PreC*(og_i) the precondition of argument og_i.

```
determine-branches (plan, I):
    PostC(plan) := UI := I;
    UPreC := ∅;
    loop OG_current := OG_last to OG_first                                /* backward trace*/
        determine-precondition (OG_current);
        if direct-contribution (OG_current, UI) then
            UI := (UI / PostC(OG_current));                               /* intended manipulation */
            UPreC := (UPreC / PostC(OG_current)) ∧ Prec(OG_current);
        elseif indirect-contribution (OG_current, UpreC) then ;
            generate-branch(OG_current);                                  /* conditioned manipulation */
                (body branch) := OG_current;                              /* generate branch */
                BC := (equal (PostC(OG_current) ∩ UPrec) nil);
                UPreC := (UPreC ∨ ((UPreC / PostC(OG_current)) ∧ Prec(OG_current)));
        else                                                              /* superfluous manipulation */
            dump (OG_current);
    end-loop;
    Prec(plan) := UPreC;
    return (plan);
end determine-branches.
```

For each Object Group checked during the backward trace three possible reasons for its occurrence exist:

1. The OG was *intended*. This means to perform the related manipulation is crucial for the achievement of the intended goal of the demonstration, namely the previously acquired user intention.
2. The OG was *not intended* but is nevertheless crucial for the achievement of the desired goal, since it enables an intended OG following later on.
3. The OG is *superfluous* and can be omitted without thwarting the achievement of the desired goal configuration.

OGs of the first kind are called to be *directly contributing* to the demonstrations, i. e. the user's goal. To test whether an Object Group is directly contributing its postcondition is simply checked against the set of conditions that represent the *unsatisfied intentions (UI)*. If there is an intersection the OG creates parts of the user intended goal state—it is directly contributing.

The second type of OGs is called to be *indirectly contributing* to the demonstration. Here, the postcondition has an empty intersection set with the unsatisfied user intentions, but there is an intersection with the set representing the *unsatisfied preconditions (UPreC)* of the following OGs. The term "intersection" used in the last phrase is meant in the way that parts of the OGs postcondition enable the positive evaluation of parts of the UPreCs. For example, we assume that as a result of a manipulation free space is provided to be occupied by another object. This might enable a manipulation that requires a subpart of this space for delivering some object. Here the amount and location of newly provided space might not be equivalent to the required space but nevertheless there is a dependency; an intersection.

The last type of OGs is fairly easy to handle. It simply does not contribute to the intended goal at all, neither directly nor indirectly.

Now the three different types of Object Groups are defined and it has been shown how a particular OG can be classified as one of these types. The remaining question is: how is an OG of a particular type to be treated during the performed backward trace in order to optimize or generalize the plan?

As outlined in the algorithm given above, a directly contributing OG is kept without any changes. Its contribution is deleted from the unsatisfied intentions and its precondition is added to the unsatisfied preconditions. Furthermore, possible indirect contributions are deleted from the UPreCs. In the contrary, OGs that are not contributing at all are simply deleted. This is in fact an optimization since an unnecessary motion sequence is erased. The interesting OGs in terms of generalization are indirectly contributing ones. Each of those is embedded in a newly generated `branch`. The *branching condition (BC)* is instantiated with an OGs indirect contribution, namely the intersection of the OGs postcondition with the unsatisfied preconditions. The unsatisfied intentions remain unchanged because there is no direct contribution. However, the unsatisfied preconditions are updated by a disjunction. One disjunctive term being the unchanged former UPreCs, corresponding to the case when the `branch` is not carried out. The other disjunctive term being the former UPreCs with the OGs indirect contribution deleted and the precondition of the `branchs` OG added. This corresponds to the case in which the manipulations embedded in the `branch` are carried out.

Backtracing through the whole demonstration, generating `branches` when possible, and updating the unsatisfied preconditions leads to a generalized and optimized plan. Its precondition which determines its applicability is equivalent to the set of accumulated unsatisfied preconditions. Obviously the "factor" of generalization corresponds directly to the number of disjunctive terms in the precondition of the resulting plan, which depends on the number of conditioned manipulations that where detected throughout the backward trace.

Further Generalization; Object Selection Criteria & Variables

Two methods exist for further user guided generalizations. The first one consists of editing the selection criteria that are determining which objects are chosen for manipulation when applying a plan. Initially an *object selection criterion (OSC)* is given from the analyzer module for every `grip` operation and therewith for every OG. This initial OSC is a conjunction of the gripping width and the position the manipulator's *Tool Center Point* had at the time the `grip` took place. Based on this initial OSC and the world model, the induction module determines the manipulated object. As soon as the object is determined the OSC is expanded with all relations of the object held at the time it was `gripped`. This includes the objects type (e. g. `cube`, `cylinder`, . . .) as well as all valid inter object relations (e. g. `on`, `in`, . . .). Thereafter the mostly constrained OSC possible is given for every OG. Based on user interaction the constraints are smoothed by introducing variables and deleting parts (literals) of the OSC. Backpropagating these changes to the plans precondition results in a generalization, widening the applicability of the plan.

Moreover, variables can be introduced by user interaction in the postcondition of the plan which reflects the users intention. For example, we assume that an object O was put into another one (O_c) which is described as being of type *case*. Furthermore these facts shall be represented in the user's intention as the relations `in`(O,O_c) and `type`(O_c,*case*). Introducing a variable for the type *case* and backpropagating this change throughout the whole plan widens its applicability since every object able to serve as a container for O can be used to instantiate O_c, not only one of type *case*.

Finally, the induction component stores the generated plan in a knowledge base. In the storing process all objects are replaced by variables. This enables the unification process necessary for plan selection and instantiation as described in section 3.3 below. Thus, generated plans can be applied to given tasks by the system's execution components. The representation of whole plans as STRIPS operators having pre- and postcondition also allows to use them as macro operators in a planning process. This way more complex tasks can be solved by combining plans that where generated from different demonstrations. All steps described so far are shown schematically in the left part of fig. 5.

3.3. Execution Phase

The third processing phase (depicted in the right part of fig. 5) addresses the problem how to select an applicable plan to a given task among previously generated ones, how to instantiate

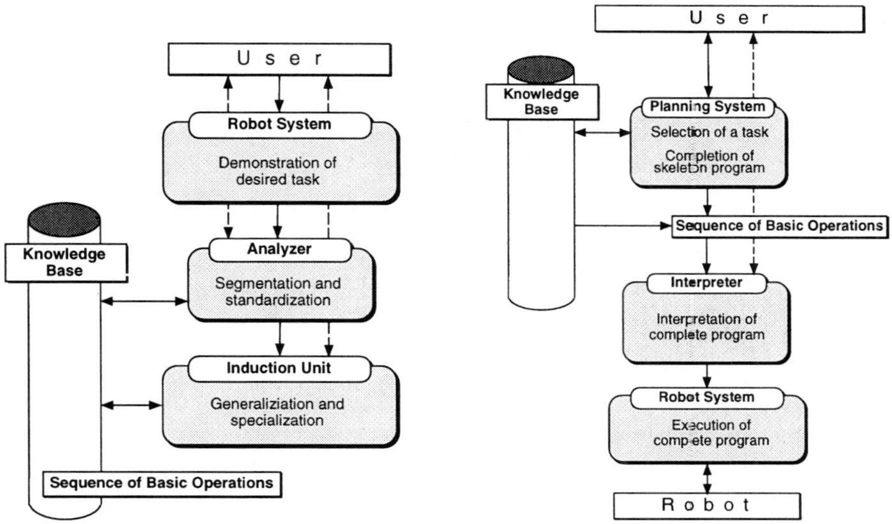

Figure 5. Phase 1 (demonstration, analysis, and induction) and Phase 2 (completion and execution) of the RPD process. Dashed arrows indicate the integration of the user.

its variables with respect to the current world model, and finally how to execute it with a given manipulator.

Plan Selection

For the selection of a plan that enables the robot system to solve a given task two prerequisites have to be fulfilled: (i) the task's goal specification has to be given as a combination of the predefined relations in *disjunctive normal form (DNF)*; (ii) the current environmental configuration has to be given as a valid world model. Given these two inputs a plan is selected by firstly matching the tasks goal specification with the postconditions of all available plans. All plans that are left after this process serve as candidates for the second selection phase. In this phase, each plan's precondition is matched with the given world model. Plans that "survive" this phase are applicable and do lead to the desired goal. Currently the first plan found to fulfill both selection phases is chosen for execution.

However, this is a greedy method leading mostly to a suboptimal selection. Future extensions will be added to assess applicable appropriate plans in terms of execution time, complexity of required motions etc. to enable a better selection.

Preparing the Execution; Plan Instantiation and Sequencing

At first glance the problem of how to prepare the selected plan for execution, namely how to instantiate its free variables and how to select the appropriate objects in the current world configuration, seems to be difficult to solve. Surprisingly, to determine the instantiations is not difficult at all. In fact, these instantiations are provided as a side effect of the selection phase. The matching processes required for plan selection include unifications of the objects in the world model with the variables in the plan's conditions, followed by the evaluation of these instantiated conditions on the world model. Therefore the determination of an applicable and appropriate plan includes one or more sets of possible and valid instantiations from which a random one is chosen. As in the selection process, the optimization of the selection of a valid instantiation set will be a future system's extension.

Based on the world model, the instantiated plan is transformed into a sequence of Basic Operations, i. e. branching conditions are tested and the bodies of the corresponding `branches` are either omitted or prepared for execution.

Solving the Task; Plan Execution

The last step is the execution of the generated program. Usually, the same robot that was used for the demonstration will be used for the execution, too. But, due to the generality of the program, it is also possible to execute the instantiated plan with another robot or with a simulation system (see fig. 5, right). Of course, the basic capabilities (degrees of freedom, sensory equipment, work space) of the system used for the execution must not differ from those supported by the system used for the demonstration. In our system, the generated program that consists of a set of basic operations can either be transformed into VAL II commands (the programming language of the controller used with our PUMA 260b) or into commands suitable for a graphical simulation interface. A transformation into other robot programming or simulation system languages is no principle problem, since only the module performing the transformation from the planning BOs to the desired language has to be added.

4. Interaction with the user—an example

In this section, the process of programming a task by demonstration, employing the methods described in section 3, is illustrated by following an example through the phases of *demonstration* and *induction*. The example is demonstrated and the generated program executed with a real industrial robot, a PUMA 260b. Thereby the system overcomes one of the most criticized aspects of former *Robot PbD systems* which have solved problems in simulated environments only (Segre, 1988) or have treated the robot as a dimension-less point in space (Andrea, 1985).

Setting up the Scene; The Example

The example task chosen for illustrating the methods used in the developed PbD system is shown in fig. 6. Two cylinders shall be placed in two designated compartments of a case.

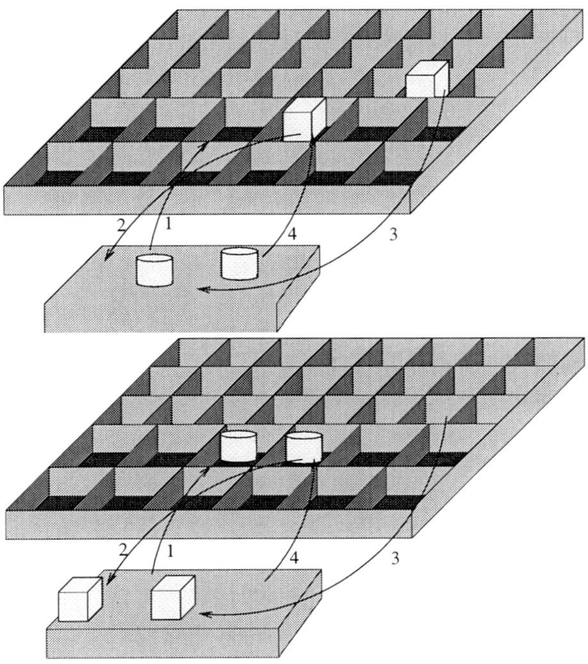

Figure 6. The start (left) and end (right) configuration of the experiment described in this section. The numbers on the arrows indicate the order in which the manipulations were performed.

As indicated by the arrows in fig. 6, the user performs four manipulations while demonstrating the task. Two of those are required in any case, namely the manipulations of the cylinders (arrows 1 and 4). The manipulation indicated by arrow 2 is required due to the given start configuration but would be obsolete if the compartment was empty from the beginning. Therefore, a branch has to be detected and generated during the induction process. The 3rd manipulation is concerned with moving the second cube needlessly with respect to the task. This motion sequence has to be eliminated later on during the induction process.

Demonstration Phase

The data recorded during the demonstration are the position of the tool center point (TCP) of the robot, the distance between the gripper jaws, and the values of the 6D bowl moved by the user. More data available like forces, torques, and orientations are not taken into account in this example. A snapshot of the raw trace recorded during the demonstration

((0 0 0 0)	(219.86 222.71 428.43)	(4939))	
((0 0 -5 0)	(219.76 222.55 429.18)	(4939))	
((0 0 0 0)	(219.76 222.55 429.18)	(4939))	
((0 0 0 2)	(219.74 222.51 429.18)	(2692))	⇒ 0 0 0 2 0
((0 0 0 0)	(219.74 222.51 429.18)	(2692))	
((0 0 40 0)	(219.79 222.74 424.78)	(2692))	⇒ 0 0 4 0 4
((0 0 42 0)	(219.96 223.07 394.19)	(2692))	⇒ 0 0 31 0 31
((-1 0 0 0)	(219.86 253.47 375.73)	(2692))	⇒ 0 30 18 0 35
((0 0 37 0)	(219.15 265.87 362.32)	(2692))	⇒ 1 13 14 0 19
((0 0 37 0)	(219.23 265.68 336.63)	(2692))	⇒ 0 0 25 0 25
((0 0 0 0)	(219.34 266.01 320.16)	(2692))	⇒ 0 0 17 0 17
((0 0 0 0)	(219.36 266.08 320.13)	(2692))	
((0 0 0 0)	(219.37 266.11 320.10)	(2692))	

Figure 7. Recorded trace data containing raw unfiltered data. Left part (from left to right): position & gripper values of the 6D bowl, position of the TCP, and gripper width. Right part: $\Delta(x_t, x_{t-1})$, gripper event & distance(x_t, x_{t-1}).

and its filtered representation is shown in fig. 7. The sequence shows a `grip` and part of a `depart` operation.

As can be seen from the data in fig. 7 (left side), sometimes the operator does not move the robot's arm at all. These standstills or "zero-movements" are eliminated by the filtering process whose parameters can be modified by the user if desired.

After the trace has been filtered, the TDNNs are applied to segment the trace and classify the segments into the five classes of *Basic Operations*. Part of the result of this step is shown in fig. 8.

(0 2)	⇒ transfer	((220, 92, 378), (220, 223, 402))
(1 14)	⇒ approach	((220, 223, 402), (220, 223, 429))
(3 17)	⇒ grip	((220, 223, 429), (2692))
(2 21)	⇒ depart	((220, 223, 425), (202, 255, 320))
(0 2)	⇒ transfer	((202, 255, 320), (-69, 270, 334))

Figure 8. Analysis results employing TDNNs; from left to right: BO identifier, start vector in filtered trace, and symbolic BO representation. `grip` & `depart` correspond to the data in fig. 7.

After the analysis the trace is presented to the user by a graphical tool. This allows to check the results of the previously performed automatic segmentation and classification. Figure 9 shows a part of the example trace. For illustrating the accuracy of the TDNN based classification, two segments that were classified to represent `approach` BOs were highlighted. The high classification accuracy can be evaluated by comparing the `approach` segments with the lightest curve which represents the Δz-values of the trace. This shows the high performance of the automatic analysis since an `approach` operation in the experiment is mainly characterized by negative translational Δz-values of the TCP.

The *demonstration* phase ends with the storage of the determined BO sequence (fig. 8).

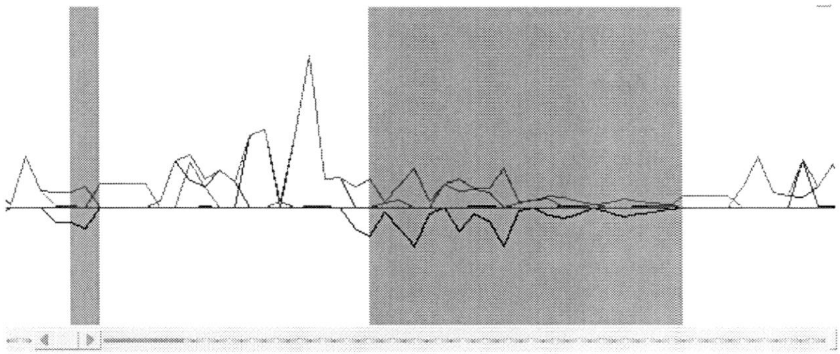

Figure 9. A snapshot of the example trace, showing translational Δx, Δy, and Δz (darkest curve) TCP values, gripper events, and two `approach` segments (grey).

Induction Phase

As described in section 3, Object Groups are formed by the induction module and the demonstrations postcondition is determined (see fig. 10).

(position(cyl1,(-68 240 530)) ∧ in(cyl1,shelf9) ∧ type(cyl1,cylinder) ∧ space(cylinder,(220 225 460)) ∧
position(cub1,(225 205 460)) ∧ on(cub1,tray) ∧ type(cub1,cube) ∧ position(cub2,(125 245 460)) ∧
on(cub2,tray) ∧ type(cub2,cube) ∧ space(cube,(-160 130 530)) ∧ position(cyl2,(-70 180 530)) ∧
in(cyl1,shelf10) ∧ type(cyl1,cylinder) ∧ space(cylinder,(307 65 460)))

Figure 10. The determined postcondition of the demonstration.

In the next step, the user has to select the intended subparts of the postcondition determined by the system. In our example, the manipulations of the two cylinders were intended which is reflected by the specified user intention shown in fig. 11.

(position(cyl1,(-68 240 530)) ∧ in(cyl1,shelf9) ∧ type(cyl1,cylinder) ∧ position(cyl2,(-70 180 530)) ∧
in(cyl2,shelf10) ∧ type(cyl2,cylinder))

Figure 11. The subpart of the plan's postcondition reflecting the intention of the user.

Applying the generalization procedure presented in section 3 leads to the generalized plan shown in fig. 12. The superfluous manipulation of the cube has been eliminated and one branch has been generated reflecting the manipulation of the other cube, which was contributing only indirectly to the solution of the example task.

By specifying the initial positions of the two cylinders on the tray as variables, further generalization is achieved. Thus the precondition of the final plan shown in fig. 13 determines the plan to be applicable when there are two cylinders on the tray, there is space in

```
OG1                           ;; put first cylinder in the shelf
If (equal (space(cylinder, (-70 180 530))) nil)
      OG2                     ;; manipulate cube if necessary
OG4                           ;; put second cylinder in the shelf
```

Figure 12. The generalized plan (manipulations are abbreviated by their Object Groups).

the first target shelf, and either also space in the second target shelf or a cube to be removed from it.

((on(cyl1,tray) ∧ type(cyl1,cylinder) ∧ space(cylinder,(-68 240 530)) ∧ on(cyl2,tray) ∧ type(cyl2,cylinder) ∧ space(cylinder,(-70 180 530)))
∨
(on(cyl1,tray) ∧ type(cyl1,cylinder) ∧ space(cylinder,(-68 240 530)) ∧ on(cyl2,tray) ∧ type(cyl2,cylinder) ∧ in(cub1,shelf10) ∧ type(cub1,cube) ∧ position(cub1,(-72 170 530)) ∧ space(cub1,(225 205 460))))

Figure 13. The precondition of the generated generalized program, determining its applicability.

5. Related work

In this section, the relationship of RPD to other research work in this area will be discussed. Due to the number of existing approaches, it is neither possible to discuss every system nor to describe each selected approach in detail. Since the RPD approach presented here is more focussed on the acquisition of procedural knowledge, the work which is more related to adaptive control method as e. g. in (Liu & Asada, 1993) or (Atkeson et al., 1987) will not be reviewed. In the following, two different "classes" of RPD will be distinguished: RPD with direct repetition and RPD with generalized repetition.

5.1. RPD with direct repetition

The term *RPD with direct repetition* is expressing that the demonstrated sequence of robot actions is used to exactly repeat the same task. The generated program cannot be applied to even slightly modified products, environments, or manufacturing situations (see section 1 above and section 12.3 in (McKerrow, 1991)). In particular, RPD with direct repetition is interesting for manufacturers in order to shorten programming cycles, provided that humans could demonstrate a task simply with their hands, i. e. in the most "natural way", while video and/or laser range cameras watch the scene. Two research projects explore exactly this technique.

Kuniyoshi et al. (1994) have called their RPD approach with direct repetition *Teaching by Showing*. A human worker builds various structures with different blocks using her or his hands while being observed by a stereo vision system. With several interpretation modules (like the movement detector, the hand tracker, the object finder, the edge junction classifier), all information necessary for the action recognizer are extracted from the image sequences. The recognizer generates a script-like action model that symbolically describes the extracted motion sequence. The recognition is model-based, i. e., it uses predefined action templates for large transfer motions as well as local motions like *approach* and *depart* movements. The recognizer instantiates those templates with the names of the identified objects, their coordinates, and spatial relation information (like alignments between objects). Finally, LISP operators like `place-on-table`, `pick` or `reach-with-obj` are created which can be sent to the robot control in order to repeat the task.

Since the RPD approach presented within this article does not use a vision system but only sequences of position tuples coming from the 6D mouse's interface, the system cannot profit from the data that an object recognizer delivers, or from spatial relations between objects detected by the edge junction classifier. Therefore, more information has to be provided to it beforehand, especially the initial position of the objects in the scene. Furthermore, the dialog-based approach for segmenting the flat sequence of positions is partly equivalent to the functionality of Kuniyoshi's change detector that identifies the "significant" changes in a scene to derive an object movement. If the segmentation mechanism fails to produce the action sequence intended, Teaching by Showing cannot correctly derive the intended code. Unlike this, our RPD system enables the demonstrator to correct the outcome of the segmentation mechanism using the RPD analyzing tool (see fig. 9). The strength of the RPD approach, however, is the induction component that allows the user to apply the generated program to many different problems of a certain class (see section 5.2). Another advantage is the capability to recognize control structures like loops (under development) and branches which is useful for real industrial tasks like loading pallets.

Research similar to that of Kuniyoshi et al. is performed by Ikeuchi and Suehiro (Ikeuchi et al., 1993). They call their approach *Assembly Plan from Observation* or *APO*. In contrast to Kuniyoshi, they use a video camera and a laser range finder. Furthermore, they have defined a more elaborated and *complete* model. The basic idea in APO is the detection of the so-called *face contact relations* between all objects and use them to derive unambiguous interpretations and generalizations of EOs. Since the model has to be able to completely explain every possible action only small problems can be solved with the APO approach.

5.2. RPD with generalized repetition

RPD with the possibility of generalizing the demonstrated programs has two advantages over RPD with direct repetition. First, RPD with generalized repetition could shorten programming times if large repetitive robot tasks like loading a pallet can be programmed by detecting loops from few example demonstrations of the repetitive steps. Second, programs with generalized repetitions can easily be adapted to new situations and, therefore, are interesting for the aspect of programming reusable code.

An application of RPD with generalized repetition was reported by R. Heise (1989) from the University of Calgary. Her system ETAR (Example-Based Task Acquisition in Robots) is able to generate programs for blocksworld constructions and, more recently, also programs for placing objects on pallets or a conveyor belt (Heise, 1992). The programs contain variables, loops, and branches. Similar to our demonstration interface, Heise also does not operate with complex vision sensors but uses the master-slave robot setup EXCALIBUR that records 10 robot locations per second.

In order to extract the important data from the stream of robot locations, Heise defined a heuristic called "focus of attention" (FOA). This focus contains all location data in the vicinity of the objects to be manipulated (vicinity is the distance of the robot gripper to an object, a configurable value, typically three inch). All location data on a path between two vicinities are thrown away and replaced by one `transport-motion` command. Although the FOA criterion considers the context of actions to segment traces, it does not guarantee a user-intended segmentation. Wrong segmentations lead to problems during the derivation of a generalized program structure because the slightly different action sequences might not match. The RPD system presented here addresses this problem by enabling additional interaction with the RPD analyzing tool (see fig. 9) to correct the segmentation if necessary.

Generalization in ETAR is realized in two steps. In the first step, the "outer generalization" derives the program structure by grouping actions and searching for loops. Like in fig. 4, subsequences enclosed by gripper operations are identified in the action sequence. In each case, they define one semantical unit that describe the transfer of objects within the scene. After the subsequences have been identified, similar subsequences will be used to induce loops. Each loop generalizes a set of similar subsequences. The method used is similar to those applied in (Dufay & Latombe, 1984) or (Andrea, 1985). Structures within the subsequences are determined using the same method and induce further inner loops or branches. Unlike the RPD system, ETAR provides no mechanism to deal with ambiguities while clustering similar subsequences. Therefore, the derived program structure might not meet the user's intention. The RPD system is using interaction (see section 4) with the user to ensure that the user-intended loops and branches are induced. The only other approaches to use additional interactions to ensure a program structure that meets the user's intention are implemented as part of the PRODEGE+ graphics editor (Sassin, 1994) or METAMOUSE (Maulsby & Witten, 1993).

In the second step, ETAR performs the "inner induction" that generalizes the parameters of the EOs as well as the determination of the loop and branch conditions. This is done using an expensive generate&test algorithm which determines exactly one solution. Nevertheless, this method does not ensure that the solution meets the user's intention. The RPD system presented does not only provide automated inner induction, but supports these steps by user interaction. This is similar to approaches in the graphics editor domain (Maulsby & Witten, 1993; Cypher, 1991; Sassin, 1994) or learning agents (Bocionek & Sassin, 1993; Bocionek, 1995). It results in safer robot programs that more probably meet the user's intention. Furthermore, the RPD approach can be applied to a larger variety of problems because it is not restricted to a fixed set of basic functions as this is necessary for ETAR's inner induction.

Another possible approach to cope with very few examples is explanation-based learning (EBL). It was explored by Segre in his ARMS system (Acquiring Robotic Manufacturing Schemata), a graphical simulation testbed for blocksworld and some peg-in-hole problems (Segre, 1988). ARMS uses EBL to derive a function schema from one example that allows him the application to situations similar with respect to the object types and spatial relations, but different with respect to geometric locations and distances. ARMS is based on a sophisticated geometric reasoner that analyzes the start and goal states of a certain assembly.

6. Conclusions and future work

In (Milne, 1994), Milne uses the metaphor of two mountaineering pioneers: Both tried to climb a mountain. One uses a horizontal and the other one a vertical approach. The first one made more progress initially, but never reached the top because there was no easy way up. The second one had much more problems in the beginning, decided to use better tools and to improve his skills, and eventually made it to the top.

In contrast to many approaches in artificial intelligence, machine learning, and robot programming by demonstration, the system presented in this paper follows the "vertical" approach. A real robot with real sensory equipment is used for the demonstration of new examples as well as for the execution of the generated programs. The system architecture (as presented in fig. 2) and the knowledge representation in the presented planning language are both very modular and can easily be extended to enhance the system's functionalities. Extensions that are currently under development and possible directions of future research will be outlined below.

6.1. Summary

In this article, an application of *Programming by Demonstration (PbD)* in the robotics domain was presented. The objective of this research is the exploration of possibilities to use machine learning for automated or semi-automated robot programming. The need for such programming environments actually comes from the industrial demand to lower the costs of production processes. With respect to the robot programming task within such processes the costs can be reduced significantly by incorporating PbD through:

- shortening programming and test cycles,
- enabling workers without or with little programming skills to perform the programming, and
- making robot programs (or program fragments) reusable for various applications.

In order to support this effort we equipped an industrial PUMA 260b manipulator with some sensors and a 6D bowl input device. Using this demonstrational interface all robot locations as well as some sensor data are recorded. The trace is first filtered to get rid of unnecessary intermediate locations. Then a segmentation takes place to find out meaningful subsequences which are classified into five classes of symbolic *Basic Operations (BOs)*. For solving this segmentation and classification task on the continuous demonstration data, subsymbolic machine learning techniques (*Time Delay Neural Networks*) are used.

After the transformation of the demonstration representation on the symbolic level the induction component is started to generalize and optimize the determined sequence of Basic Operations. Therefore, the planning language contains operations for `branches` and `loops` besides the BOs, all of those modeled as STRIPS-like operators. Moreover a variable concept is integrated. The postcondition for each operator and the whole demonstration is determined. By incorporating a man-machine interface, the user is asked questions about the intentions related to the demonstration. By analyzing the demonstration with respect to the user's intentions superfluous and conditioned subparts are detected. Superfluous manipulations are eliminated, whereas `branches` are generated whenever conditioned manipulations are encountered. With this method, supported by introducing variables by the user in a dialogue, optimization and generalization takes place. The resulting plan is stored for later usage in a knowledge base.

The execution module selects an appropriate and applicable plan to a given task and environment from the available ones. The selected plan is prepared for execution by instantiating its variables with respect to the given environmental configuration and is transformed into a sequence of executable operations. Finally, the prepared plan is translated to the specific robot control language of the desired robot system and executed.

6.2. Discussion

Although the automatic analysis of the recorded demonstration traces and the induction component's optimization and generalization modules perform very good, some potential weaknesses of the RPD system have to be discussed within this section.

First of all, the lack of external sensors limits the overall system performance. It must fully rely on the closed-world assumption introduced by STRIPS, is not able to autonomously build up and maintain a world model, and cannot execute a plan reactively. As outlined in the following section, external sensor data will be integrated in the PbD process to overcome this weakness.

Another potential drawback of the system is the use of STRIPS-like operators in the planning language. One might ask: "What is the benefit of RPD when the STRIPS operators, a world model, and a goal specification are already available? Simply apply a standard planning algorithm, and sooner or later it will produce a plan that solves the task!" Although this is true in principle, planning is an NP-hard problem which means that—given a sufficiently complex world model and goal specification—the off line planning will take *very* long. Therefore, in many cases it is much more comfortable and faster for the user to give a demonstration. Moreover, it is also reasonable to integrate a standard planner into the RPD system, i. e. to combine PbD and off line planning in order to make the best use of both methods.

Finally, the fact that the system's performance relies fundamentally on the successful user interaction might be seen as a drawback. Undoubtedly, a system which is able to learn and generalize autonomously by always guessing the user's intention right would be the ultimate PbD system. Unfortunately, a system with such capabilities will not be implemented in the near future. Today's ML algorithms often learn plausible concepts with respect to the input data that nevertheless do not reflect the user's intention. Obviously, this behavior cannot be tolerated for robot programming systems that shall be applied in manufacturing or service domains. Therefore, the tight but comfortable integration of the user in the demonstration process supports safe control and is rather a necessity and advantage than a weakness of the presented system.

6.3. Future Research

To extend the presented approach to more complex applications including interactions of grasped parts with other objects is the challenge for future work. Particular extensions are:

- *Reactive behavior:* Until now, the robot can operate in a static, well-structured and well-known environment only where nothing unforeseen happens, because no external sensors are used. In the future, an on line planner and an exception handler which provide reactive behavior to the robot will be added to the system. Therefore, external sensors that permanently observe the robot's world, e. g. vision sensors, are going to be used.
- *New applications:* If the robot is able to react to specific actions in the world by using external sensors, new applications become possible, e. g. manufacturing and assembling operations. Moreover, new operators representing pre-learned low-level skills (see Baroglio et al., 1996; Kaiser et al., 1995) will be integrated in the planning language. This will enlarge the basic manipulation capabilities of the system allowing compliant motions, e. g. insertion operations.
- *Induction of loops:* Some of the systems presented in section 5 do support the detection of repetitively occurring demonstration parts and to induce loop structures from them. An extension that will enable the system presented here to induce `loop` structures, too, is currently under development. Again, the focus has been set on the generation of user intended `loops`, i. e. `loops` that do not only reflect a repetitively occurring structure, but also the user's intention.

Adding these extensions to the RPD system will improve its applicability and planning performance significantly. Finally, the result will be a system which might serve as a programming system in real-world domains like manufacturing or service environments (e. g. households etc.), supporting the user in programming even complex tasks easy and comfortable by demonstration.

Acknowledgments

The RPD prototype system has been developed and realized at the Institute for Real-Time Computer Systems and Robotics, Prof. Dr.-Ing. U. Rembold and Prof. Dr.-Ing. R. Dillmann, Department of Computer Science, University of Karlsruhe. We gratefully acknowledge the support of Jürgen Kreuziger and Michael Kaiser, who helped to develop the RPD system with many ideas and fruitful discussions. A special thanks to Michael Kaiser for implementing and maintaining the robot's interface. We would also like to thank the unknown reviewers of this article who made invaluable suggestions to help us revise the first submission.

Notes

1. Despite these problems, *teaching by showing* is still the most popular method in industrial applications today, especially in mass production.
2. Noisy compared to how the motion vector sequence of the robot would look like if the motions would be executed automatically.
3. Object Groups are closely related to the Key Groups introduced in (Heise, 1989).

References

Andreae, P. M. (1985). Justified Generalization: Acquiring Procedures from Examples. *Technical Report AI-TR-834*. Artificial Intelligence Laboratory, MIT.

Atkeson, C. G., Aboaf, E. W., McIntyre, J., & Reinkensmeyer, D. J. (1987). Model-based robot learning. In *Proceedings of the 4th International Symposium on Robotics Research*.

Baroglio, C., Giordana, A., Kaiser, M., Nuttin, M., & Piola, R. (1996). Learning Controllers for Industrial Robots. *Machine Learning, same issue*.

Bocionek, S. (1995). Agent systems that negotiate and learn. *International Journal Human-Computer Studies, 42*, 265–288.

Bocionek, S. & Sassin, M. (1993). Dialog-Based Learning (DBL) for Adaptive Interface Agents and Programming-by-Demonstration Systems. *Technical Report CMU-CS-93-175*. Carnegie Mellon University, Pittsburgh.

Cypher, A. (1991). EAGER: Programming Repetitive Tasks by Example. In *CHI '91 Conference Proceedings* (pp. 33–39). New Orleans, Louisiana: ACM Press.

Dufay, B. & Latombe, J.-C. (1984). An approach to automatic robot programming based on inductive learning. *International Journal of Robotics Research, 3*, 97–115.

Fikes, R. E. & Nilsson, N. J. (1971). Strips: A new approach to the application of theorem proving to problem solving *Artificial Intelligence, 2*, 189–208.

Flaig, T., Neugebauer, J.-G., & Wapler, M. (1994). VR4RobotS: a New Off-line Programming System Based on Virtual Reality Techniques. In *Proceedings of the 25th International Symposium on Industrial Robots* (pp. 671–678). Hannover, Germany.

Friedrich, H. & Kaiser, M. (1995). What can Robots learn from Humans?, In *IFAC Workshop on Human-Oriented Design of Advanced Robotic Systems* (pp. 1–6). Vienna, Austria.

Heise, R. (1989). Demonstration Instead of Programming: Focussing Attention in Robot Task Acquisition. *Research Report 89/360/22*. University of Calgary.

Heise, R. (1992). Programming Robots by Example. *Research Report 92/476/14*. University of Calgary.

Ikeuchi, K., Kawade, M., & Suehiro, T. (1993). Towards Assembly Plan from Observation: Task Recognition with Planar, Curved and Mechanical Contacts. In *Proceedings of the IEEE/RJS International Conference on Intelligent Robots and Systems* (pp. 2294–2301). Yokohama, Japan.

Kaiser, M., Giordana, A., & Nuttin, M. (1994). Integrated Acquisition, Execution, Evaluation and Tuning of Elementary Skills for Intelligent Robots. In *Proceedings of the IFAC Symposium on Artificial Intelligence in Real Time Control* (pp. 145–150). Valencia, Spain.

Kaiser, M., Retey, A., & Dillmann, R. (1995). Robot skill acquisition via human demonstration. In *Proceedings of the International Conference on Advanced Robotics* (pp. 763–768), Barcelona, Spain.

Kreuziger, J. & Hauser, M. (1993). A New System Architecture for Applying Symbolic Learning Techniques to Robot Manipulation. In *Proceedings of the IEEE/RSJ International Conference on Intelligent Robots and Systems*. Yokohama, Japan.

Kuniyoshi, Y., Masayuki, I., & Inoue, H. (1994). Learning by watching: Reusable task knowledge from visual observation of human performance. *IEEE Transactions on Robotics and Automation, 10*, 799–822.

Liu, S. & Asada, H. (1993). Teaching and training of deburring robots using neural networks. In *IEEE International Conference on Robotics and Automation* (pp. 339–345).

Lozano-Pérez, T. (1983). Robot Programming. In *Proceedings of the IEEE, 71*, (pp. 821–841).

Maulsby, D. L. & Witten, I. H. (1993). Metamouse: An Instructable Agent for Programming by Demonstration. In A. Cypher (Ed.), *Watch What I Do: Programming by Demonstration*. MIT Press.

McKerrow, P. J. (1991). Introduction to Robotics. In *Electronic Systems Engineering*. Addison-Wesley.

Milne, R. (1991). Building Successful Applications: The Wrong Way and the Right Way. In G. Barth et al. (Eds.) *KI-94 – Anwendungen der Künstlichen Intelligenz*. Springer.

Münch, S., Kreuziger, J., Kaiser, M. & Dillmann, R. (1994). Robot Rrogramming by Demonstration (RPD) – Using Machine Learning and User Interaction Methods for the Development of Easy and Comfortable Robot Programming Systems. In *Proceedings of the 25th International Symposium on Industrial Robots* (pp. 685–693). Hannover, Germany.

Münch, S., Sassin, M. & Bocionek, S. (1994b) The Application of PbD Methods to Real-World Domains: Two Case Studies. In *Proceedings of the 7th Australian Joint Conference on Artificial Intelligence* (pp. 92–99). Armidale, Australia.

Neubauer, W., Bocionek, S., Möller, M. & Rencken, W. (1992). Learning Systems Behavior for the Automatic Correction and Optimization of Off-line Robot Programs. In *Proceedings of the IEEE International Conference on Intelligent Robots and Systems*, Raleigh.

Sassin, M. (1994). Creating user-intended programs with programming by demonstration. In *1994 Proceedings of the IEEE/CS Symposium on Visual Languages* (pp. 153–160). St. Louis, Missouri.

Segre, A. M. (1988). *Machine Learning of Robot Assembly Plans*. Kluwer Academic Publishers.

Thrun, S. B. & Mitchell,T. M. (1993). Integrating Inductive Neural Network Learning and Explanation-based Learning. In *Proceedings of the 13th International Joint Conference on AI* (pp. 930–936). Chambery, France.

Ude, A. (1993). Trajectory Generation from Noisy Positions of Object Features for Teaching Robot Paths. *Robotics and Autonomous Systems, 11*, 113–127.

Ude, A., Bröde, H. & Dillmann, R. (1994). Object Localization Using Perceptual Organization and Structural Stereopsis. In *Proceedings of the 3rd International Conference on Automation, Robotics and Computer Vision*, Singapore.

Waibel, A., Hanazawa, T., Hinton, G., Shikano, K. & Lang, K. (1989). Phoneme recognition using time-delay neural networks. *IEEE Transactions on acoustics, speech and signal processing*, 328–339.

Machine Learning, 23, 191–220 (1996)

Performance Improvement of Robot Continuous-Path Operation through Iterative Learning Using Neural Networks

PETER C.Y. CHEN†, JAMES K. MILLS† AND KENNETH C. SMITH†‡ chencyp@me.utoronto.ca
†*Department of Mechanical Engineering, University of Toronto, 5 King's College Road, Toronto, Ontario, Canada M5S 1A4*

‡*Department of Electrical and Computer Engineering, University of Toronto, 10 King's College Road, Toronto, Ontario, Canada M5S 1A4*

Editors: Judy A. Franklin, Tom M. Mitchell, and Sebastian Thrun

Abstract. In this article, an approach to improving the performance of robot continuous-path operation is proposed. This approach utilizes a multilayer feedforward neural network to compensate for model uncertainty associated with the robotic operation. Closed-loop stability and performance are analyzed. It is shown that the closed-loop system is stable in the sense that all signals are bounded; it is further proved that the performance of the closed-loop system is improved in the sense that certain error measure of the closed-loop system decreases as the network learning process is iterated. These analytical results are confirmed by computer simulation. The effectiveness of the proposed approach is demonstrated through a laboratory experiment.

Keywords: robot control, neural networks, uncertainty compensation, stability and performance

1. Introduction

In many industrial operations, a robot is required to follow a continuous path accurately. An example of this type of operation is arc welding, where the end-effector of the robot is required to follow a prescribed path with a prescribed velocity. Currently, industrial robots for continuous-path operation (also known as trajectory tracking) are "programmed" by the so-called "walk-through" method, where an operator physically guides the robot through a sequence of desired movements which is then stored in the controller's memory. Such a programming method is time consuming and uneconomical, because during the walk-through, the robot is out of production activity.

An alternative to the walk-through method is the so-called "off-line programming" method, where a high-level programming language is used to write a control program which specifies actions of each actuator that would produce the desired motion of the robot. Currently industrial robots are PID-controlled. (A PID controller is a controller with three terms in which the output of the controller is the sum of a proportional term, an integral term, and a differentiating term, with adjustable gain for each term (Dorf, 1992).) Off-line programming for continuous-path operation based on PID control may not produce the desired robot motion for the following reason. Since the robot dynamics is nonlinear while PID control is a linear-control method, applying PID control to the trajectory tracking problem would require a gain scheduling approach using local models; that is, the robot dynamics is linearized about some operating points so that the control gains can be selected to achieve certain performance specification. The trajectory, however, varies with time (as opposed to a set-point), therefore the gains so selected may not be appropriate due to the linearization. As a result, significant tracking error may occur.

A technique known as "computed torque" (Spong & Vidyasagar, 1989) has been shown to be an effective alternative to PID control under a condition called "exact cancellation of nonlinearity". In the computed torque method, control actions are generated based on a mathematical model of the robot. Exact cancellation of nonlinearity would result in a linear decoupled system, but it requires that the parameters in the dynamics model of the robot be known exactly. Failure to meet this condition leads to certain undesirable signals called "uncertainty", which must be compensated in order to improve the performance of the robot in continuous-path operation.

Investigations dealing with uncertainty reported in the robotics literature can in principle be classified under the following two approaches: robust control and adaptive control. The premise of robust control (Corless & Letimann, 1981) is that although the uncertainty is unknown, it is possible to estimate the "worst case" bounds on its effect on the tracking performance of the manipulator. The robust control law is designed with the objective to overcome the effect of the uncertainty (rather than to "cancel" the uncertainty so that a linear decoupled system is obtained). In the adaptive control approach (Ortega & Spong, 1989), the basic premise is that by changing the values of gains or other parameters in the control law according to some on-line algorithm, the controller can find a set of values for these gains or parameters so that the trajectory tracking error is reduced. Stability analysis of these approaches often makes use of the Second Method of Lyapunov, which guarantees that the tracking error will be reduced to zero or a small neighborhood of zero as time goes to infinity. Other than in the context of exponential stability, which is much more difficult to obtain, Lyapunov stability generally provides no clear insight about the transient performance of the manipulator (Spong & Vidyasagar, 1989).

A class of computational models known as neural networks has been applied to system control in general and to robot control in particular, e.g., (Miller, Sutton & Werbos, 1990). (The use of the word "neural" to describe such computational models stems solely from modern convention. Although their structure may have been derived from neuronal models of the central nervous system, the computational models discussed in this article are, at most, only mathematical abstractions of biological neuronal systems.) Justification for using neural networks for robot control is based on the following properties of neural networks: (i) The ability of the neural network to "learn" (through a repetitive training process) (McClelland & Rumelhart, 1986) enables a controller incorporated with a neural network to improve its performance. (ii) The ability of the neural network to "generalize" what it has learned (Denker, et al., 1987) enables the controller also to respond to unexpected situations. (iii) The structure of neural networks allows massive parallel processing, especially when the neural networks are implemented in hardware using VLSI technology (Gelsner & Pöchmüller, 1994); such inherent collective processing capability enables the neural network to respond quickly in generating timely control actions.

It is due to these properties that a neural-network-based approach to uncertainty compensation could be considered potentially advantageous over both robust control and adaptive control. The learning ability of neural networks is especially desirable in controlling robots that perform repetitive manufacturing tasks. One reason for using robots instead of human workers in manufacturing is that robots can perform repetitive tasks with better quality and consistency. Unavoidable in repetitive robotic operation in an industrial setting, however, is the sustained "wear-and-tear" (e.g., joint friction, wear of gears, etc.) of the robot. Such wear-and-tear inevitably effects the dynamic characteristics of the robot. In other words, the wear-and-tear introduces uncertainty into the robotic system, and consequently degrades its performance. A neural network that learns (iteratively) to compensate for the effect of such wear-and-tear would enable the manipulator to maintain satisfactory performance consistently throughout its expected lifetime.

In this article, we propose an approach to robot trajectory tracking using a multilayer feedforward neural network. (In the sequel, the term "neural network" or just "network" is sometimes used instead of "multilayer feedforward neural network" for convenience.) The neural network is used explicitly to compensate for the uncertainty in the manipulator. We address the dynamical behavior of the manipulator in a two-step analysis. First we show that the closed-loop system based on the proposed approach is stable in the sense that all signals in the system are bounded. We then show that the neural network improves the performance of the robot in the sense that certain error measure of the closed-loop system is reduced as the learning process of the neural network is iterated. We subsequently present simulation results that confirm the analytical conclusions, and experimental results that demonstrate the effectiveness of the proposed approach.

The contributions of this work to the application of neural networks to robot control are: (i) The insight obtained (through the analysis on the dynamics of the neural network) on the stability and performance of the closed-loop system with the neural network learning on-line is significant. The results of the analysis confirm that neural networks could be used as plausible tools for robot control within the context of uncertainty compensation. (ii) The experimental implementation of the proposed approach together with the positive experimental results reported in this article clearly demonstrate the effectiveness of the neural network as an uncertainty compensator for practical robotic tasks. Many studies on the application of neural networks to system control in general and to robot control in particular rely on numerical simulation to verify the conclusions therein; very few proposed schemes have been physically implemented to verify their effectiveness. Extensive and conclusive experimental results are needed in order to affirm neural networks as viable tools for robot control. The experiment reported in this article represents an incremental step in gathering such results.

This article is organized as follows. Section 2 reviews the literature on the application of neural networks to robot trajectory tracking. Section 3 formulates the dynamics of the robotic system and presents the control law that incorporates a compensating signal from the neural network. Section 4 shows that a compensator exists in the form of a multilayer feedforward neural network, and describes the algorithm used for neural network learning. Section 5 presents analysis of stability and performance. Section 6 describes the computer simulation conducted to verify the analytical conclusions, and presents the results. Section 7 describes the experimental implementation of the proposed approach, and presents the experimental results. Section 8 discusses the implications of the results and suggests directions for future research.

2. Literature Review

Various robot control techniques using neural networks reported in the literature can in principle be classified into two general schemes according to the role of the neural network. Figure 1 illustrates the first scheme.

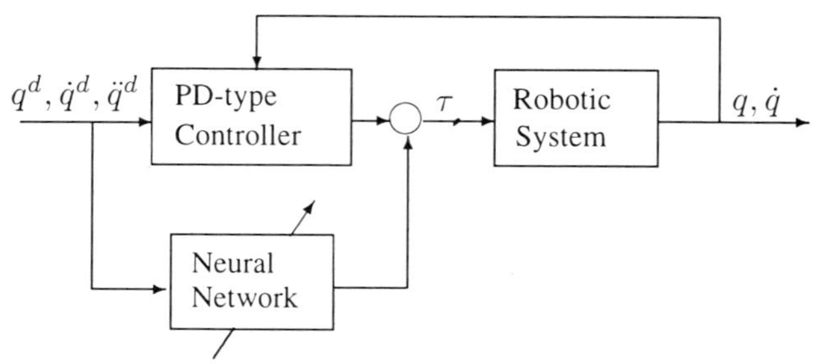

Figure 1. Neural Network as Inverse Dynamics Model.

In this first scheme, the neural network is to repetitively learn to represent the inverse dynamics of the manipulator, so that by feeding the desired trajectory into the neural network, the desired torque signal is produced as the output of the network. The PD controller is used mainly to stabilize the closed-loop system and to guide the repetitive learning process of the neural network. (A PD controller is a controller with two terms in which the output of the controller is the sum of a proportional term and a differentiating term, with adjustable gain for each term (Dorf, 1992).) Gomi and Kawato (1990) use this scheme for robot trajectory tracking, and present computer simulation results involving a two-link manipulator. Cılız and Işık (1990) utilize this scheme to control a manipulator under payload variation. Yabuta and Yamada (1991) apply this scheme (without the PD controller) to manipulator control, and demonstrate its effectiveness using an one degree-of-freedom servomechanism. Arai, Rong and Fukuda (1993) study the possibility of increasing the speed of the learning process.

In the second scheme, illustrated in Figure 2, the neural network is used to deal with uncertainty in the model parameters. The inverse dynamics control law is used to generate an approximate torque signal. This torque signal is then augmented by a compensating signal generated by a neural network. The neural network is to learn to generate the proper compensating signal by adjusting its weights so as to maximize some performance measure, such as reduction of the tracking error. Okuma and Ishiguro (1990) apply this scheme to the control of a manipulator with consideration of joint friction, and present results of computer simulation involving a two-link manipulator. Kuan and Bavarian (1992) also study, again through computer simulation, the problem of dealing with joint friction using this scheme. Zomaya and Nabhan (1993) apply this scheme, without the PD controller, to control a manipulator, and conduct computer simulation using the dynamics model of a $PUMA$ 560 robot.

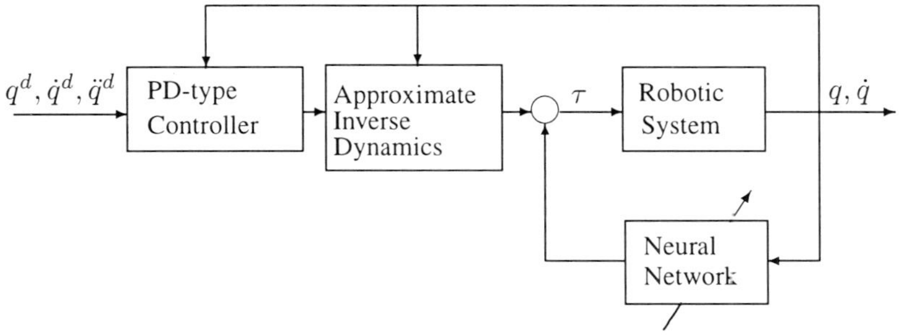

Figure 2. Neural Network as Compensator.

Although attempts have been made, among the studies cited above, to analyze the stability and performance of these schemes, the results reported remain inconclusive, mainly because they are obtained based on very restrictive assumptions. It appears that in the current literature on the application of neural networks to robot control, the issues of closed-loop stability and performance have not been resolved. The importance of resolving these issues is underscored by the fact that, as part of the control system, the dynamics of the neural network is coupled with that of the manipulator; consequently, the dynamical behavior of the neural network, namely the process of weight adjustment, will inevitably affect the stability as well as the performance of the closed-loop system.

It also appears that many studies on the application of neural networks to robot control rely on numerical simulation to verify the conclusions therein; very few proposed schemes have been physically implemented to verify their effectiveness.

It is in the context of these observations that the work described in this article complements other studies reported in the literature. In this article, we show that the closed-loop system based on the proposed approach is stable and that the neural network improves the performance of the closed-loop system through iterative learning. Our analytical results confirm that neural networks can be used as effective tools to improve the performance of a manipulator in continuous-path operation. We subsequently present the results of the experimental implementation of our proposed approach involving a laboratory manipulator. The experimental results clearly demonstrate the effectiveness of the neural network in improving the performance of the robotic system. In short, the work presented in this article complements other published studies both in analytical aspect and in experimental aspect.

3. Dynamics and Control

In general, the dynamics of a manipulator with n joints can be described by a set of nonlinear differential equations, compactly expressed in the form (Spong & Vidyasagar, 1989)

$$M(q)\ddot{q} + h(q, \dot{q}) = \tau, \tag{1}$$

where $q \in \mathcal{R}^n, \dot{q} \in \mathcal{R}^n$, and $\ddot{q} \in \mathcal{R}^n$ are respectively the joint position, joint velocity, and joint acceleration vectors, $M(\cdot) \in \mathcal{R}^{n\times n}$ is the inertia matrix, $h(\cdot) \in \mathcal{R}^n$ is a vector containing the Coriolis, gravitational, centrifugal and frictional terms, and $\tau \in \mathcal{R}^n$ is the input torque vector. With regard to an industrial robot, the following characteristics usually apply: (i) the manipulator is composed of serial links; (ii) the manipulator is not redundant; and (iii) the links and joints of the manipulator are rigid.

Supposing that the terms $M(\cdot)$ and $h(\cdot)$ are known precisely, the inverse dynamics control law can be written as

$$\tau = M(q)u + h(q, \dot{q}), \tag{2}$$

where u is a control input to be specified. Substituting (2) into (1) yields: $\ddot{q} = u$. Let u be a PD-type control of the form

$$u = \ddot{q}^d + K_v(\dot{q}^d - \dot{q}) + K_p(q^d - q), \tag{3}$$

where $q^d \in \mathcal{R}^n$, $\dot{q}^d \in \mathcal{R}^n$, and $\ddot{q}^d \in \mathcal{R}^n$ are respectively the desired joint position, joint velocity, and joint acceleration vectors, and $K_v \in \mathcal{R}^{n\times n}$ and $K_p \in \mathcal{R}^{n\times n}$ are diagonal gain matrices. Let $e = \begin{bmatrix} q^d - q \\ \dot{q}^d - \dot{q} \end{bmatrix}$ and $A = \begin{bmatrix} 0 & I \\ -K_p & -K_v \end{bmatrix}$. Then the error dynamics of the closed-loop system can be expressed as

$$\dot{e} = Ae, \tag{4}$$

which represents a linear decoupled system whose dynamical behavior can be completely specified by selecting appropriate values for the gains K_v and K_p (Dorf, 1992).

The inverse dynamics control law (2) is idealized because it results in a linear decoupled system only if the the values of the parameters in $M(\cdot)$ and $h(\cdot)$ are known precisely. In practice, such precise knowledge about a physical system is not available. Thus the realistic inverse dynamics control law takes the form

$$\tau = \hat{M}(q)u + \hat{h}(q, \dot{q}), \tag{5}$$

where $\hat{M}(\cdot)$ and $\hat{h}(\cdot)$ are respectively the estimates of $M(\cdot)$ and $h(\cdot)$. Now with (3) and (5), the resulting closed-loop error dynamics becomes

$$\dot{e} = Ae + B\eta(q, \dot{q}, \ddot{q}), \tag{6}$$

where $B = [0, I]^T$, and $\eta(q, \dot{q}, \ddot{q}) = (\hat{M}^{-1}M - I)\ddot{q} + \hat{M}^{-1}(h - \hat{h})$. Note that for brevity, the arguments of M, h, $\hat{M}$, and $\hat{h}$ have been omitted.

The system (6) is exactly the same as the linear decoupled system (4) if the term $\eta(\cdot)$ is identically zero. This term exists whenever $\hat{M}(\cdot) \neq M$ and/or $\hat{h}(\cdot) \neq h(\cdot)$, and is referred to as the *uncertainty*. Due to the presence of η, the dynamical behavior of the system (6) can no longer be "shaped" as desired by simply selecting appropriate values for the gains K_v and K_p.

To reduce the effect of the uncertainty, a compensating signal v is introduced as part of the control signal u as follows

$$u = \ddot{q}^d + K_v(\dot{q}^d - \dot{q}) + K_p(q^d - q) + v. \tag{7}$$

Substituting (5) and (7) into (1) yields the error dynamics of the closed-loop system

$$\dot{e} = Ae + B\Delta v, \tag{8}$$

where $\Delta v = \eta(q, \dot{q}, \ddot{q}) - v$ is referred to as the *control error*. If v is generated such that $\Delta v \to 0$, then (8) becomes: $\dot{e} = Ae$, which is again the linear decoupled system (4). The benefit gained through uncertainty compensation (i.e., $\Delta v \to 0$) is that, because the robotic system is rendered linear and decoupled even in the presence of the uncertainty η, the dynamics of the robotic system can now be controlled to meet specific performance requirements by selecting appropriate values for the gains. Figure 3 schematically depicts the closed-loop system.

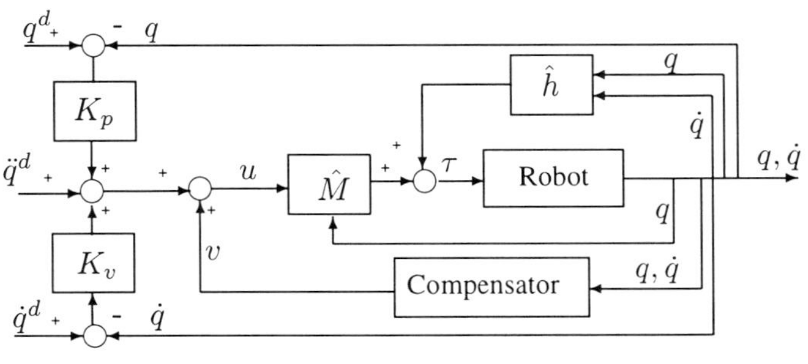

Figure 3. Uncertainty Compensation in Robot Control.

4. Uncertainty Compensation

The objective of uncertainty compensation is to generate the appropriate compensating signal v such that the control error Δv vanishes, i.e., $\Delta v \to 0$. An ideal compensator is a function whose output v exactly equals that of the function $\eta(\cdot)$. Based on such a premise, the problem of uncertainty compensation can then be considered as a function approximation problem. A multilayer feedforward neural network represents an attractive mechanism for dealing with such a problem, because it is capable of approximating any continuous function, and more importantly, it can *learn* to approximate any given continuous function (McClelland & Rumelhart, 1986).

A multilayer feedforward neural network consists of a collection of processing elements (or units) arranged in a layered structure (McClelland & Rumelhart, 1986) as shown in Figure 4.

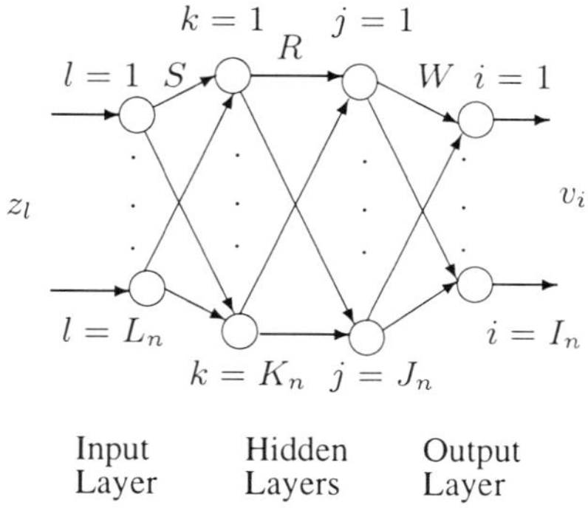

Figure 4. Neural Network Structure.

For the neural network with two hidden layers, as depicted in Figure 4, the mapping realized by the network can be described as follows. Let the number of units in the input layer, the first hidden layer, the second hidden layer, and the output layer be L_n, K_n, J_n, and I_n respectively. Let $\bar{\bar{r}}_k, \bar{r}_j$, and r_i represent the input to the first hidden layer, the second hidden layer, and the output layer respectively, and let $\bar{\bar{v}}_k, \bar{v}_j$ and v_i represent the corresponding output of these layers, then $\bar{\bar{r}}_k = \sum_{l=1}^{L_n} S_{kl} z_l$, $\bar{\bar{v}}_k = g(\bar{\bar{r}}_k)$, $\bar{r}_j = \sum_{k=1}^{K_n} R_{jk} \bar{\bar{v}}_k$, $\bar{v}_j = g(\bar{r}_j)$, $r_i = \sum_{j=1}^{J_n} W_{ij} \bar{v}_j$, and $v_i = g(r_i)$, where $g(x) = c\tanh(x)$, and c is a scaling factor. For convenience a generalized weight vector Θ is defined as: $\Theta = [W_1, .., W_i, .., W_{I_n}, R_1, .., R_j, .., R_{J_n}, S_1, .., S_k, .., S_{K_n}]$, where $(\cdot)_a$ represents the ath row of the matrix $(\cdot)$. The mapping realized by the network can then be compactly expressed as: $v = N(Z, \Theta)$, where Z is the input vector, i.e. $Z = (z_1, z_2, ...z_l, ...z_{L_n})$, and N is used as a convenient notation to represent the mapping achieved by the network.

It has been proven that a multilayer feedforward neural network with one hidden layer (containing a sufficient number of units) is capable of approximating any continuous function to any degree of accuracy (Cybenko, 1989). It is in this sense that multilayer feedforward neural networks have been established as a class of "universal approximators". Thus, for the uncertainty function $\eta(Z)$, where $Z = (q, \dot{q}, \ddot{q})$, there exists a set of weights Θ^* for a neural network (with a sufficient number of hidden units) with the output $v^d = N(Z, \Theta^*)$ such that, for some $\epsilon \geq 0$, $\|\eta - v^d\| \leq \epsilon$, where $\|(\cdot)\|$ denotes the supremum of $(\cdot)$.

Note that the above statement only assures that the weights Θ^* exist, it does not indicate what their values are, or how to find them. To determine these weights is the objective of neural network learning. Let the cost function to be minimized be: $J_{\Delta v} = \frac{1}{2}\Delta v^T \Delta v$.

Applying the error-backpropagation algorithm (McClelland & Rumelhart, 1986) yields: $\dot{\Theta} = -\lambda_n \frac{\partial J_{\Delta v}}{\partial \Theta} = -\lambda_n \Delta v^T \frac{\partial \Delta v}{\partial \Theta}$, where λ_n is the learning rate. Since $\Delta v = v^d - v$, where v^d is the unknown desired output of the network, and $\frac{\partial v^d}{\partial \Theta} = 0$, the weight update rule becomes: $\dot{\Theta} = \lambda_n \Delta v^T \frac{\partial v}{\partial \Theta}$. Specifically, the dynamics of the weights W_{ij}, R_{jk}, and S_{kl} can be expressed (Muller & Reinhardt, 1990) as: $\dot{W}_{ij} = \lambda_n \Gamma_i \bar{v}_j$, $\dot{R}_{jk} = \lambda_n \bar{\Gamma}_j \bar{\bar{v}}_k$, and $\dot{S}_{kl} = \lambda_n \bar{\bar{\Gamma}}_k z_l$, where $\Gamma_i = \Delta v_i g'(v_i)$, $\bar{\Gamma}_j = g'(\bar{v}_j) \sum_{i=1}^{I_n} \Gamma_i W_{ij}$, $\bar{\bar{\Gamma}}_k = g'(\bar{\bar{v}}_k) \sum_{j=1}^{J_n} \bar{\Gamma}_j R_{jk}$, and $g'(\cdot) = \frac{\partial g(\cdot)}{\partial(\cdot)}$. The learning process is illustrated in Figure 5.

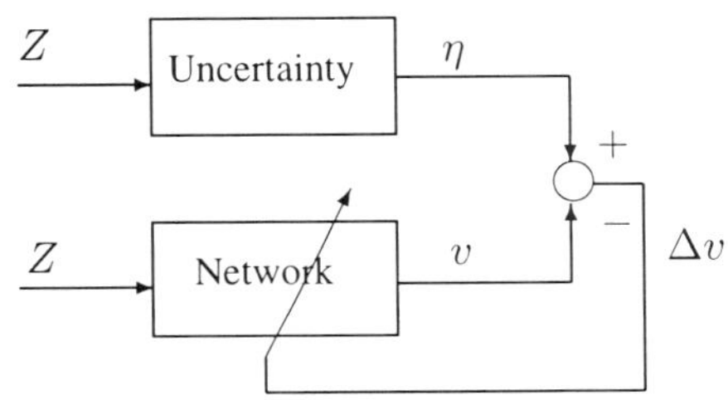

Figure 5. Neural Network for Uncertainty Compensation.

The control error Δv can be determined in real-time, according to (8), as follows

$$\Delta v = \ddot{q}^d - \ddot{q} + K_v(\dot{q}^d - \dot{q}) + K_p(q^d - q). \tag{9}$$

The closed-loop dynamics of the robotic system with the neural network learning on-line is described by

$$\begin{cases} \dot{e} = Ae + B\Delta v(q, \dot{q}, \ddot{q}, \Theta) \\ \dot{\Theta} = -\lambda_n \Delta v^T(q, \dot{q}, \ddot{q}, \Theta) \frac{\partial \Delta v(q, \dot{q}, \ddot{q}, \Theta)}{\partial \Theta}. \end{cases} \tag{10}$$

Stability and performance of the system (10) are examined next.

5. Analysis

We first prove that the closed-loop system with the neural network learning on-line is stable in the sense that all signals in the system are bounded. We then show that the performance of the system is improved in the sense that certain measure of the control error Δv decreases as the learning process is iterated (i.e., as the number of learning trials increases). The conjecture is that reduction in the control error eventually leads to reduction in the trajectory tracking error. This conjecture is verified by the results of the computer simulation presented in Section 6.

5.1. Stability

THEOREM 1 *Given a continuous and twice-differentiable reference position trajectory $q^d(t)$, the system (10) is bounded-input bounded-output (BIBO) stable for sufficiently large gains K_v and K_p.*

Note that a system is said to be BIBO stable if for a bounded input, the output of the system is also bounded. This term is defined rigorously in Appendix A. Proof of the above theorem is presented in Appendix B. It suffices to state here that this theorem asserts that for sufficiently large gains, the control input u and the trajectory tracking error e in (10) are bounded.

COROLLARY 1 *The acceleration signal $\ddot{q}(t)$ is bounded.*

COROLLARY 2 *The control error Δv is bounded.*

COROLLARY 3 *The weights of the neural network remain bounded during a given trial.*

Proofs for Corollaries 1, 2, and 3 can be found in Appendix C, D, and E respectively. We next examine how the network learning process affects the performance of the manipulator.

5.2. Performance Improvement

Preliminaries

Let t represent the continuous time variable, i.e., $0 \leq t < \infty$. Let learning start at time $t = 0$, and let each trial last T seconds. Then the pth trial spans the time period from $t = (p-1)T$ to $t = pT$. Note that p is thus implicitly defined as a positive integer.

Let ξ be the time variable associated with one trial, i.e., $0 \leq \xi \leq T$. (The notation $\dot{x}$ from here on means either $\frac{dx}{dt}$ or $\frac{dx}{d\xi}$ as it should be clear from the context.) Let $x(p,\xi)$ denote the value of the variable x at the ξth second of the pth trial. Then $x(p,0)$ represents the value of the variable x at the beginning of the pth trial, and $x(p,T)$ represents the value of the variable x at the end of the pth trial. Note that $x(p,0) = x(p-1,T)$. Let Θ_m be an element of the weight vector Θ. Let $\Delta\theta_m(p,\xi)$ denote the change of Θ_m during the first ξ seconds of the pth trial, i.e., $\Delta\theta_m(p,\xi) = \int_0^\xi \dot{\Theta}_m(p,\sigma)d\sigma$.

The L_∞^n-norm of a Lebesgue integrable function $f(t) : R_+ \rightarrow R^n$, denoted by $\|f\|_\infty$, is defined as: $\|f\|_\infty = ess\,sup_{t\in[0,\infty)}\|f(t)\| < \infty$. The *extended* L_∞^n-space (with the truncated L_∞^n-norm $\|(\cdot)_T\|_\infty$), denoted by $L_{\infty e}^n$, is defined as: $L_{\infty e}^n = \{f : R_+ \rightarrow R^n \,|\, f_T \in L_\infty^n, \forall T < \infty\}$, where $f_T(t) = \begin{cases} f(t) & \text{for } t \in [0,T] \\ 0 & \text{otherwise} \end{cases}$. For convenience, the notation $\|f\|_{T\infty}$ is used to denote $\|f_T\|_\infty$.

Let $C[0,T]$ denote the family of Lebesgue integrable function $f_i(\xi)$ for all $\xi \in [0,T]$. The L_2-norm of a vector function $f(\xi) = (f_1, f_2, ..., f_i, ..., f_n), f_i \in C[0,T]$, over the time interval $[0,T]$ is defined (Vidyasagar, 1992) as: $\|f\|_{2T} = \left(\int_0^T f^T(\xi)f(\xi)d\xi\right)^{\frac{1}{2}}$. For convenience, the notation $\|\cdot\|$ instead of $\|\cdot\|_{2T}$ is used to denote this norm in the subsequent analysis.

PROPOSITION 1 *For the closed-loop system (10), there exists some small $\lambda_n > 0$ such that $\|\Delta\theta_m(p+1,\xi) - \Delta\theta_m(p,\xi)\|_{T\infty} << 1$, for all $\xi \in [0,T]$.*

The proof of this proposition is presented in Appendix F.

Remark: Since $\Delta\theta_m(p+1,\xi)$ represents the amount of weight change during the first ξ seconds of the $(p+1)^{th}$ trial, and $\Delta\theta_m(p,\xi)$ represents the amount of weight change during the first ξ seconds of the pth trial, this proposition (illustrated in Figure 6) means that, for $\lambda_n << 1$, the difference between the change in Θ_m of any two successive trials can be considered negligible, i.e., $\Delta\theta_m(p+1,\xi) \approx \Delta\theta_m(p,\xi), \forall\xi \in [0,T]$.

A qualitative interpretation for this proposition can be constructed based on observation on the time-scale difference between the dynamics of the manipulator and the dynamics of the on-line learning process of the neural network. With a small learning rate λ_n, the overall robotic system can be considered as a two-time-scale system with the manipulator exhibiting a "fast" dynamics while the network exhibiting a "slow" dynamics. As the learning rate λ_n approaches zero, the change in the weights *per trial* can be expected to be infinitesimally small. Such small change in the weights will not have significant effect on the state of the robot. Therefore, between any two *successive* trials, the change in the state of the robot, and consequently the difference between the two amounts of weight change, can be considered negligible. This proposition is verified by computer simulation presented in Section 6.

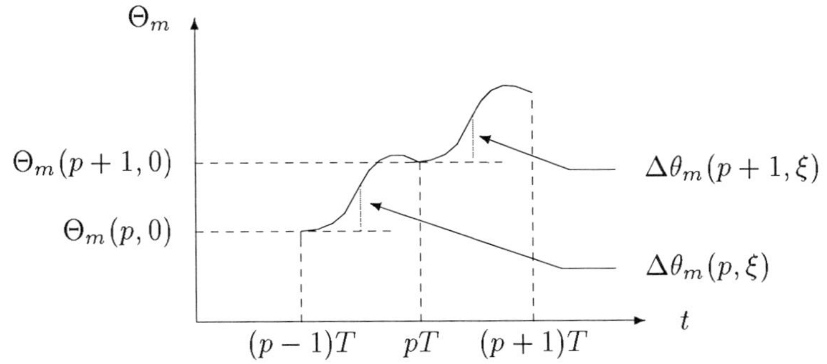

Figure 6. Weight and Weight Change.

Main Results

Recall from Section 4 that the objective of neural network learning is to reduce the control error Δv. The following theorem establishes rigorously that such an objective can be achieved.

THEOREM 2 *The L_2-norm of the control error Δv decreases as the number of trials p increases, i.e., $\|\Delta v(p+1)\| < \|\Delta v(p)\|$.*

Proof: Since (from Section 4) $J_{\Delta v}(p,\xi) = \frac{1}{2}\Delta v^T(p,\xi)\Delta v(p,\xi)$, from the definition of the L_2-norm, we have $\|\Delta v(p)\|^2 = 2\int_0^T J_{\Delta v}(p,\xi)d\xi$. To prove that $\|\Delta v(p+1)\| < \|\Delta v(p)\|$, it suffices to show that $\|\Delta v(p+1)\|^2 - \|\Delta v(p)\|^2 < 0$. Now $\|\Delta v(p+1)\|^2 - \|\Delta v(p)\|^2 = 2\int_0^T (J_{\Delta v}(p+1,\xi) - J_{\Delta v}(p,\xi))\, d\xi$. Based on the fact that the change in the control error Δv between any two successive trials p and $(p+1)$ is a direct consequence of the change in the network weights, we expand $J_{\Delta v}(p+1,\xi)$ about $J_{\Delta v}(p,\xi)$, while ignoring the higher order terms, to obtain

$$J_{\Delta v}(p+1,\xi) - J_{\Delta v}(p,\xi) \approx \sum_{m=1}^{c_\theta} \left[\frac{\partial J_{\Delta v}}{\partial \Theta_m}\left(\Theta_m(p+1,\xi) - \Theta_m(p,\xi)\right)\right], \qquad (11)$$

where c_θ is the total number of weights, i.e., $c_\theta = I_n \times J_n + J_n \times K_n + K_n \times L_n$. Because $\frac{\partial J_{\Delta v}(p,\xi)}{\partial \Theta_m(p,\xi)} = \frac{\partial J_{\Delta v}(p,\xi)}{\partial \Delta v(p,\xi)}\frac{\partial \Delta v(p,\xi)}{\partial \Theta_m(p,\xi)} = \Delta v^T(p,\xi)\frac{\partial \Delta v(p,\xi)}{\partial \Theta_m(p,\xi)}$, so (11) becomes

$$\begin{aligned} J_{\Delta v}(p+1,\xi) &- J_{\Delta v}(p,\xi) \\ &= \sum_{m=1}^{c_\theta}\left[\Delta v^T(p,\xi)\frac{\partial \Delta v(p,\xi)}{\partial \Theta_m(p,\xi)}\left(\Theta_m(p+1,\xi) - \Theta_m(p,\xi)\right)\right]. \qquad (12)\end{aligned}$$

Note that $\Theta_m(p+1,\xi)$ and $\Theta_m(p,\xi)$ can be expressed as $\Theta_m(p+1,\xi) = \Theta_m(p+1,0) + \Delta\theta_m(p+1,\xi)$, and $\Theta_m(p,\xi) = \Theta_m(p,0) + \Delta\theta_m(p,\xi)$, where $\Theta_m(p+1,0) = \Theta_m(p,T)$ represents the weight value at the end of the pth trial, while $\Theta_m(p,0)$ represents the weight value at the beginning of the pth trial. From Proposition 1, we have $\|\Delta\theta_m(p+1,\xi) - \Delta\theta_m(p,\xi)\|_{T\infty} \ll 1$, hence $\Theta_m(p+1,\xi) - \Theta_m(p,\xi) \approx \Theta_m(p+1,0) - \Theta_m(p,0)$, and consequently

$$\begin{aligned} &\|\Delta v(p+1)\|^2 - \|\Delta v(p)\|^2 \\ &= 2\int_0^T \left(J_{\Delta v}(p+1,\xi) - J_{\Delta v}(p,\xi)\right) d\xi \\ &= 2\int_0^T \sum_{m=1}^{c_\theta}\left[\Delta v^T(p,\xi)\frac{\partial \Delta v(p,\xi)}{\partial \Theta_m(p,\xi)}(\Theta_m(p+1,0) - \Theta_m(p,0))\right] d\xi \\ &= 2\sum_{m=1}^{c_\theta}\left[\int_0^T \Delta v^T(p,\xi)\frac{\partial \Delta v(p,\xi)}{\partial \Theta_m(p,\xi)}(\Theta_m(p+1,0) - \Theta_m(p,0))d\xi\right]. \qquad (13)\end{aligned}$$

Since $\Theta_m(p+1,0)$ and $\Theta_m(p,0)$ are no longer functions of ξ, so (13) becomes

$$\begin{aligned} &\|\Delta v(p+1)\|^2 - \|\Delta v(p)\|^2 \\ &= 2\sum_{m=1}^{c_\theta}\left[\left(\int_0^T \Delta v^T(p,\xi)\frac{\partial \Delta v(p,\xi)}{\partial \Theta_m(p,\xi)}d\xi\right)(\Theta_m(p+1,0) - \Theta_m(p,0))\right].\end{aligned}$$

Note that $\Theta_m(p+1,0) - \Theta_m(p,0) = \int_0^T \dot{\Theta}_m(p,\xi)d\xi$. But from the learning rule presented in Section 4, we have $\dot{\Theta}_m(p,\xi) = -\lambda_n \Delta v^T(p,\xi)\frac{\partial \Delta v(p,\xi)}{\partial \Theta_m(p,\xi)}$. Thus $\Theta_m(p+1,0) -$

$\Theta_m(p,0) = \int_0^T \dot{\Theta}_m(p,\xi)d\xi = -\lambda_n \int_0^T \Delta v^T(p,\xi)\frac{\partial \Delta v(p,\xi)}{\partial \Theta_m(p,\xi)}d\xi$. Consequently,

$$\begin{aligned}
&\|\Delta v(p+1)\|^2 - \|\Delta v(p)\|^2 \\
&= 2\sum_{m=1}^{c_\theta}\left[\left(\int_0^T \Delta v^T(p,\xi)\frac{\partial \Delta v(p,\xi)}{\partial \Theta_m(p,\xi)}d\xi\right)\left(-\lambda_n \int_0^T \Delta v^T(p,\xi)\frac{\partial \Delta v(p,\xi)}{\partial \Theta_m(p,\xi)}d\xi\right)\right] \\
&= -2\lambda_n \sum_{m=1}^{c_\theta}\left[\left(\int_0^T \Delta v^T(p,\xi)\frac{\partial \Delta v(p,\xi)}{\partial \Theta_m(p,\xi)}d\xi\right)^2\right] \\
&\leq 0.
\end{aligned}$$

Now $\|\Delta v(p+1)\|^2 - \|\Delta v(p)\|^2 = 0$ if $\sum_{m=1}^{c_\theta}\int_0^T \Delta v^T(p,\xi)\frac{\partial \Delta v(p,\xi)}{\partial \Theta_m(p,\xi)}d\xi = 0$. But this implies that $\int_0^T \dot{\Theta}_m d\xi = 0$, which states that the total change in the weight Θ_m for the trial p is zero. This means that the gradient search conducted by the error-backpropagation algorithm has reached either a global minimum or a local minimum. If this is not the case, then we have $\|\Delta v(p+1)\|^2 - \|\Delta v(p)\|^2 < 0$, that is, $\|\Delta v(p+1)\| < \|\Delta v(p)\|$, which implies that the controller error Δv decreases as the number of trials p increases.

■

Remark: An important practical issue concerning neural network learning is whether the weights will remain bounded. This issue can be resolved (in the context of the above analysis) by observing the facts that (i) the weights are bounded during a given trial (i.e., Corollary 3), and (ii) from an implementation standpoint, the learning process can be terminated once the L_2-norm of the control error no longer decreases from trial to trial. Thus, if we start the learning process with a set of finite weights, and if the weight dynamics condition (i.e., Proposition 1) is satisfied, then it is assured that the weight values are finite at the point when the learning process is terminated.

6. Simulation

The purpose of conducting computer simulation is to verify the analytical results presented in Section 5. Specifically, we conducted the simulation to confirm, through a numerical example, that (i) for a small learning rate λ_n, the proposition that the weight change between two successive trials can be considered negligible is valid, (ii) upon confirmation of (i), the L_2-norm of the control error Δv decreases as the number of learning trial p increases, and (iii) reduction in the control error Δv results in reduction in trajectory tracking error e.

The manipulator considered in the simulation, as shown in Figure 7, consists of two links with point masses; that is the mass of each link is assumed to be concentrated at the tip of each link.

The dynamics of this manipulator can be formulated (Craig, 1986) as

$$\tau_1 = m_2 l_2^2(\ddot{q}_1 + \ddot{q}_2) + m_2 l_1 l_2(2\ddot{q}_1 + \ddot{q}_2)\cos q_2 + (m_1 + m_2) l_1^2 \ddot{q}_1$$

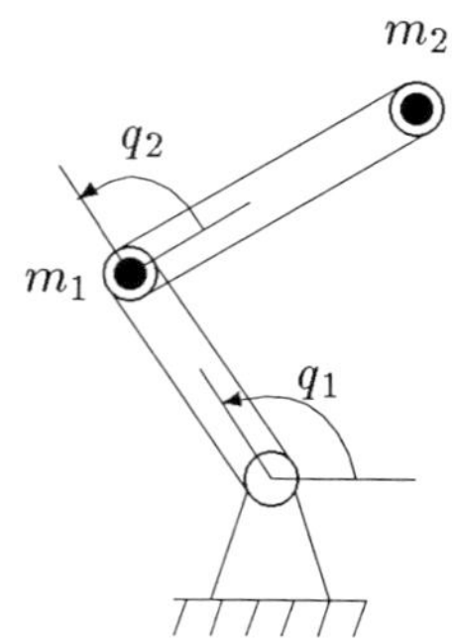

Figure 7. A Two-Link Manipulator.

$$
\begin{aligned}
& -m_2 l_1 l_2 \dot{q}_1^2 \sin q_2 - 2m_2 l_1 l_2 \dot{q}_1 \dot{q}_2 \sin q_2 + m_2 l_2 g \cos(q_1 + q_2) \\
& +(m_1 + m_2) l_1 g \cos q_1, \\
\tau_2 \;=\; & m_2 l_1 l_2 \ddot{q}_1 \cos q_2 + m_2 l_1 l_2 \dot{q}_1^2 \sin q_2 + m_2 l_2 g \cos(q_1 + q_2) + m_2 l_2^2 (\ddot{q}_1 + \ddot{q}_2),
\end{aligned}
$$

where q_1 and q_2 are the angles of the two joints (as indicated in Figure 7) with the corresponding torques τ_1 and τ_2, m_1 and m_2 are the mass of the links, l_1 and l_2 are the length of the links, and $g = 9.8 \;\; m/s^2$. To simulate the dynamics of the manipulator, the following parameter values were used: $m_1 = 5.0 \; kg$, $m_2 = 4.0 \; kg$, $l_1 = 0.7 \; m$, and $l_2 = 0.5 \; m$. In implementing the inverse dynamics control, the values for the masses were altered to introduce uncertainty into the system by setting: $\hat{m}_1 = 4.5 \; kg$ and $\hat{m}_2 = 5.0 \; kg$. The desired position trajectory for both joints, generated by a 5th-order polynomial, is as shown in Figure 8.

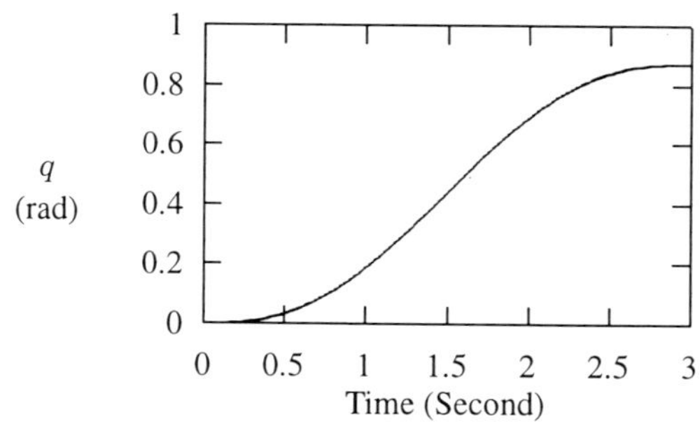

Figure 8. Desired Position Trajectory.

A network consisting of six input units, two hidden layers with twenty units in each layer, and two output units was used in the simulation. The learning λ_n was set at a very small value of 5×10^{-5}, so as to be consistent with the argument for Proposition 1 discussed in Section 5. The initial weights of the neural network for the first trial were set to arbitrary values in the order of 10^{-4}. The control gains were set to be: $K_p = \text{diag}[30, 30]$ and $K_v = \text{diag}[10, 10]$. A series of simulation runs (i.e., trials) were carried out. The control error Δv was calculated according to (9), which requires the acceleration signal $\ddot{q}$. In the simulation, $\ddot{q}$ was estimated on-line based on $\dot{q}$.

To examine the dynamical behavior of the weights between two successive trials, Figures 9 and 10 show respectively the dynamics and the difference between the change of a connection weight $R_{(10,10)}$ (the weight between the 10th unit of the first hidden layer and the 10th unit of the second hidden layer) during the 10th trial and the 11th trial.

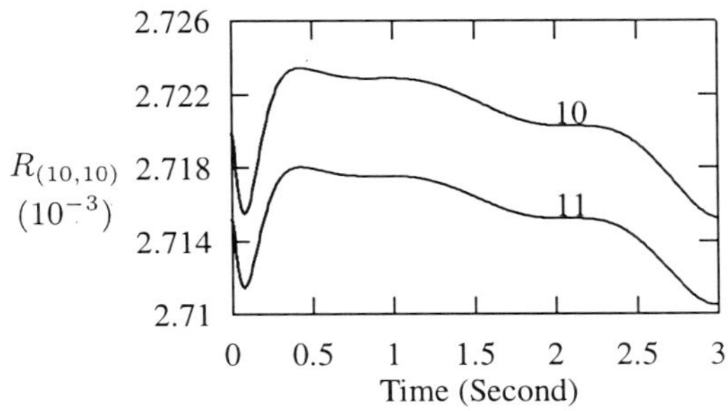

Figure 9. Dynamics of Weight $R_{(10,10)}$: 10th and 11th Trial.

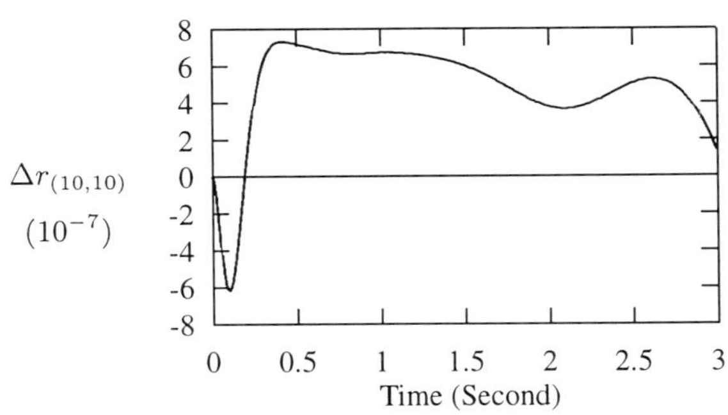

Figure 10. Difference between Change of Weight $R_{(10,10)}$: 10th and 11th Trial.

It is shown here, as an example, how the difference between the change in $R_{(10,10)}$ of the two trials, 10th and 11th, is calculated. Using the notations defined in Section 5, the difference of weight change between trial 10 and trial 11 can be written as: $\Delta r_{(10,10)}(\xi) = \Delta R_{(10,10)}(11,\xi) - \Delta R_{(10,10)}(10,\xi)$, where $\Delta R_{(10,10)}(11,\xi) = R_{(10,10)}(11,\xi) - R_{(10,10)}(11,0)$, $\Delta R_{(10,10)}(10,\xi) = R_{(10,10)}(10,\xi) - R_{(10,10)}(10,0)$, and ξ is the time variable, i.e., $\xi \in [0,3]$.

Figures 11 and 12 show respectively the dynamics and the difference between the change of the same weight during the 20th trial and the 21st trial. From Figures 9-12, it can be seen that the difference of the weight change between two successive trials (i.e., 10th and 11th, 20th and 21st) is indeed negligible.

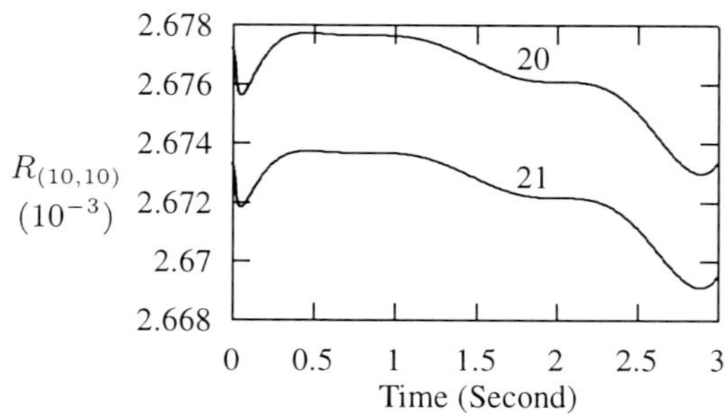

Figure 11. Dynamics of Weight $R_{(10,10)}$: 20th and 21st Trial.

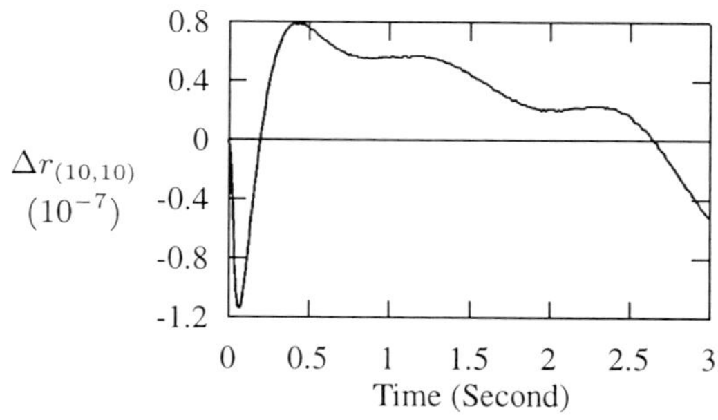

Figure 12. Difference between Change of Weight $R_{(10,10)}$: 20th and 21st Trial.

Figures 13-16 show the dynamics and the difference between the change of another weight $W_{(1,10)}$ (the weight between the 10th unit of the second hidden layer and the 1st unit of the output layer) during the same pair of trials (i.e., 10th and 11th, 20th and 21st).

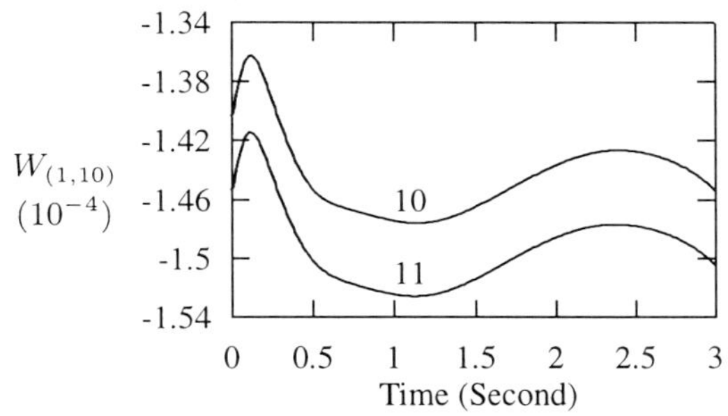

Figure 13. Dynamics of Weight $W_{(1,10)}$: 10th and 11th Trial.

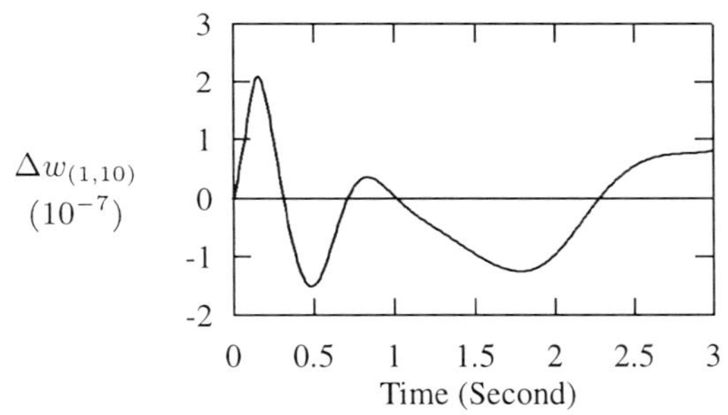

Figure 14. Difference between Change of Weight $W_{(1,10)}$: 10th and 11th Trial.

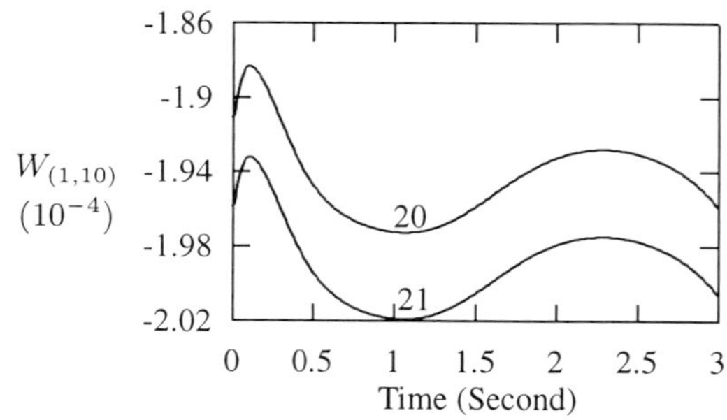

Figure 15. Dynamics of Weight $W_{(1,10)}$: 20th and 21st Trial.

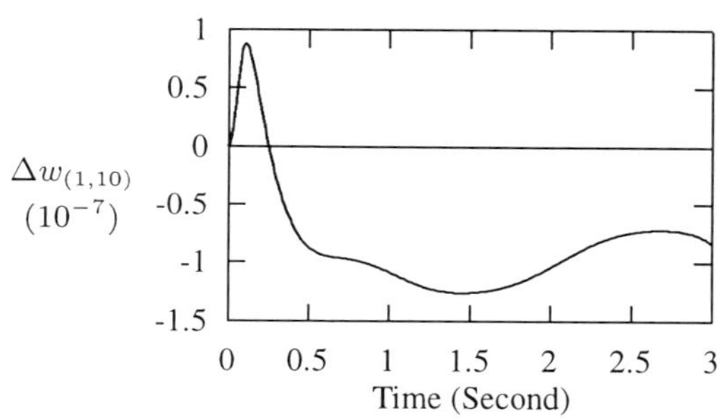

Figure 16. Difference between Change of Weight $W_{(1,10)}$: 20th and 21st Trial.

Comparing Figures 9, 11, 13, and 15 with Figure 6 (which illustrates the proposition that the difference between the weight change of any two successive trials is negligible), it is evident that, when the learning rate is small, the neural network does indeed possess the dynamical behavior as predicted.

Recall that the analytical conclusion in Section 5 based on Proposition 1 is that the L_2-norm of the control error Δv decreases as the number of trials p increases. Figure 17 shows the L_2-norm of the control error Δv versus the trial number p for this simulation.

It can be seen that the control error Δv indeed decreases as the number of trials p increases. This confirms the theoretical conclusion presented in Section 5. It is conjectured that reduction in the control error Δv eventually results in reduction in the trajectory tracking error. Figures 18 and 19 show the joint errors without compensation (i.e., $v = 0$) and during the 300th trial. It can be seen that the joint errors are reduced as the learning of the network progresses.

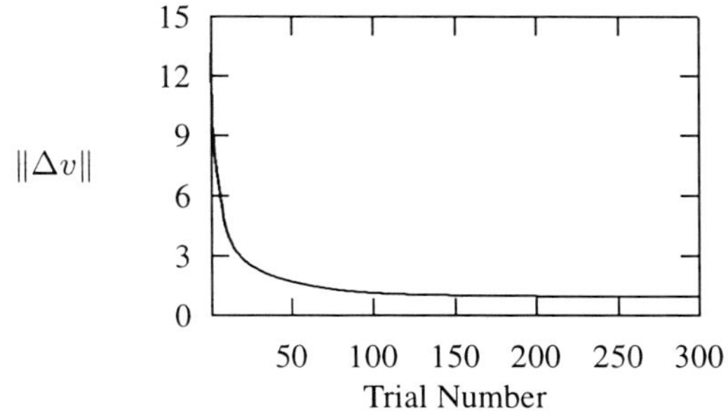

Figure 17. Control Error.

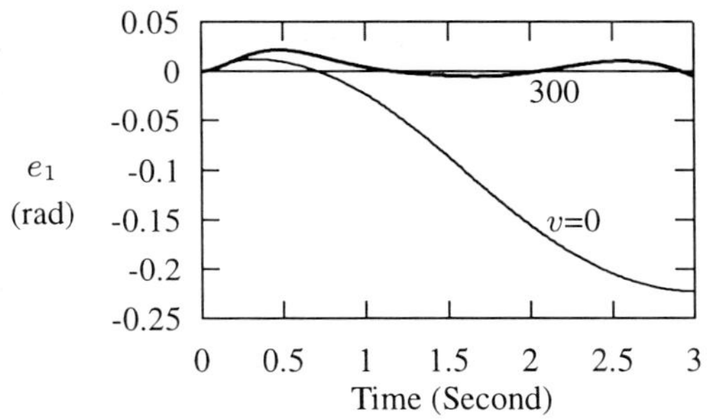

Figure 18. Error in q_1.

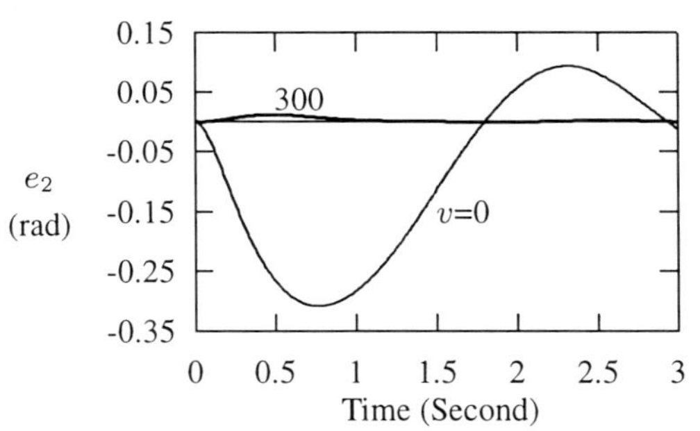

Figure 19. Error in q_2.

From the results of the simulation, it can be seen that both the proposition (that with a small learning rate, the weight change between two successive trials can be considered negligible) and the analytical conclusion (that the L_2-norm of the control error decreases as the number of trials p increases) are valid. It can also be seen that reduction in the control error Δv indeed results in reduction in the trajectory tracking error. A laboratory experiment conducted to demonstrate the effectiveness of the proposed approach is described next.

7. Experiment

The robot used for implementing the proposed approach is of a five-bar parallel link configuration operating in the horizontal plane (Lokhorst, 1990) as shown in Figure 20.

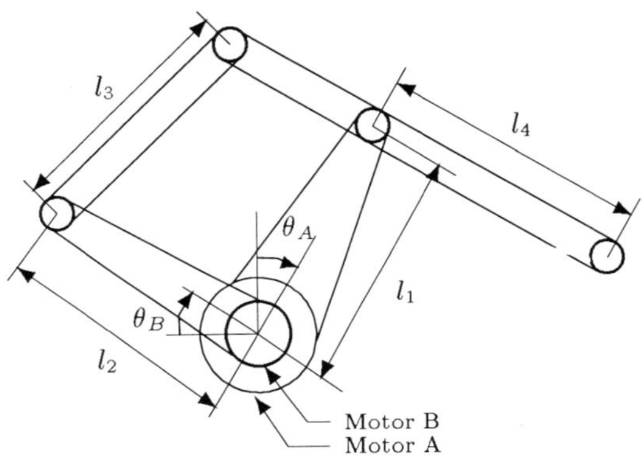

Figure 20. A Laboratory Manipulator.

High torque, brushless direct current motors are used to drive the manipulator without gear reduction. The motors used are manufactured by Yokogawa Corporation. S[illegible]ations of the motors are listed in Table 1.

Table 1. Direct-drive Motor Specification.

Specification	*Motor A*	*Motor B*
Manufacturer	Yokogawa	Yokogawa
Model	DMA 12000	DMB 1045
Maximum Torque (Nm)	200	45
Maximum Speed (rev/s)	1.2	2.4
Encoder Resolution $(pulse/rev)$	1024000	655360
Diameter (mm)	264	160
Length (mm)	188	143
Weight (kg)	29	9.5

The dynamics of this manipulator can be formulated as (Lokhorst, 1990)

$$\begin{bmatrix} d_1 & d_3 S_{BA} \\ d_3 S_{BA} & d_2 \end{bmatrix} \begin{bmatrix} \ddot{\theta}_A \\ \ddot{\theta}_B \end{bmatrix} + \begin{bmatrix} -d_3 C_{BA} \dot{\theta}_B^2 \\ d_3 C_{BA} \dot{\theta}_A^2 \end{bmatrix} + \begin{bmatrix} b_1 \dot{\theta}_A + b_3 sgn(\dot{\theta}_A) \\ b_2 \dot{\theta}_B + b_4 sgn(\dot{\theta}_B) \end{bmatrix} = \begin{bmatrix} \tau_A \\ \tau_B \end{bmatrix},$$

where $S_{BA} = \sin(\theta_B - \theta_A)$, $C_{BA} = \cos(\theta_B - \theta_A)$, θ_A and θ_B are the angles of the two motors with the corresponding torques τ_A and τ_B, $d_1 = m_1 l_{c1}^2 + m_3 l_{c3}^2 + m_4 l_1^2 + I_1 + I_3$, $d_2 = m_2 l_{c2}^2 + m_3 l_2^2 + m_4 l_{c4}^2 + I_2 + I_4$, and $d_3 = m_3 l_2 l_{c3} - m_4 l_1 l_{c4}$. Here m_i is the mass of the i-th link, l_i is the length of the i-th link, l_{ci} is the distance from the joint to the center of mass of the i-th link, and I_i is the moment of inertia of the i-th link. The lengths of the links are: $l_1 = l_3 = l_4 = 0.5\ m$, and $l_2 = 0.3\ m$. The parameters b_1 and b_2 are the coefficients of viscous friction and the parameters b_3 and b_4 are the coefficients of static friction. The values of the parameters have been experimentally determined (Lawryshyn, 1993) and are shown in Table 2 under the heading "True".

Table 2. True and Estimated Parameter Values for Robot Trajectory Tracking Experiment.

Parameter	*True*	*Estimated*
d_1	$3.4\ kgm^2$	$1.2\ kgm^2$
d_2	$-0.1\ kgm^2$	$-0.511\ kgm^2$
d_3	$1.2\ kgm^2$	$0.772\ kgm^2$
b_1	$8.3\ kgm^2/s$	$8.3\ kgm^2/s$
b_2	$1.6\ kgm^2/s$	$1.6\ kgm^2/s$
b_3	$2.3\ Nm$	$2.3\ Nm$
b_4	$0.46\ Nm$	$0.46\ Nm$

A neural network with six input units, two output units, and ten units in each of its two hidden layers was implemented in the form of a program written in the C programming language for the experiment. Due to the limitation of hardware and computational resources — the control laws were executed on a 386 microcomputer with a clock speed of $25MHz$, the learning process of the neural network was implemented off-line. This means that during each trial, only the static mapping of the network was activated, while the weights of the network were fixed. Data required for network learning were recorded during the trial. Once a trial was completed, the weights of the network were updated using the recorded data. Because network learning was conducted off-line, Theorem 2 does not apply to this experiment. The significance of the experiment lies in demonstrating the effectiveness of the proposed approach in the execution of realistic robotic tasks.

The weights of the neural network were randomly initialized to be in the order of 10^{-3}. The learning rate of the neural network, initially set at 1×10^{-3}, was gradually reduced to 5×10^{-5} as the learning process was iterated. The input signals to the neural network were $q, \dot{q}$, and $\ddot{q}$. The joint acceleration $\ddot{q}$ was obtained by filtering the the joint velocity signal $\dot{q}$ on-line using a Kalman filter (Yanovski, 1991). The feedback gains were set to be: $K_v = \text{diag}[15, 15]$ and $K_p = \text{diag}[50, 50]$. The control law was executed at a sampling frequency of $86Hz$.

A task trajectory for the tip of the manipulator as shown in Figure 21 was planned in this experiment. This trajectory was generated off-line and stored in memory, and retrieved during real-time task execution.

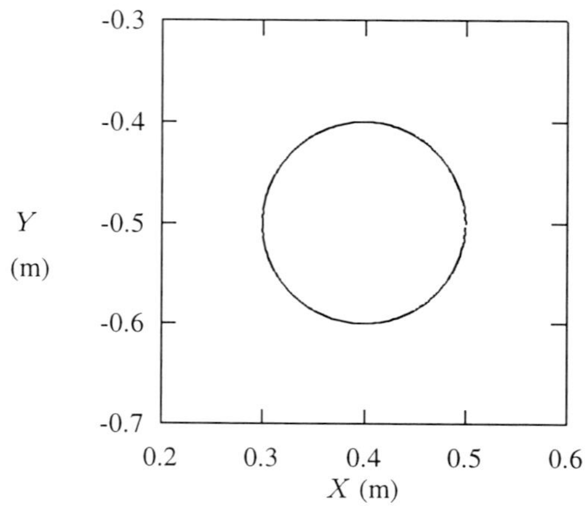

Figure 21. Desired Task Trajectory.

To establish a basis for comparison, a baseline experiment was first conducted using the "true" parameters of the manipulator as shown in Table 2. Figure 22 shows the results of this experiment. This baseline experiment shows the performance of the system when there is presumably no uncertainty in the system parameters.

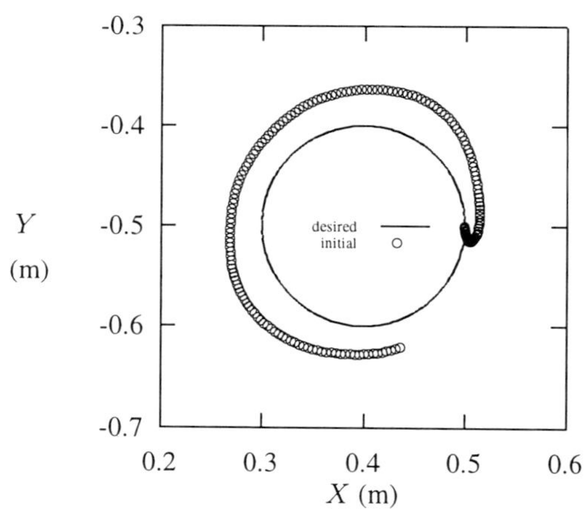

Figure 22. Desired Trajectory and Baseline Trajectory.

It is noted that even when the "true" parameter values were used in generating the control signal, the trajectory of the tip of the manipulator still deviates from the desired trajectory quite significantly. Such deviation can be accounted for by considering (i) the effect of time-delay in executing the control law with a sampling frequency of only 86 Hz, and (ii) the fact that the control gains K_v and K_p were set at low values. The reason for using such low gains in this experiment was to allow a larger initial tracking error (when uncertainty but no compensation was introduced) so that the effect of neural network learning in improving the performance of the manipulator could be more clearly seen. Since the key objective in this experiment was to show performance improvement, the emphasis was on showing reduction in the "size" of the tracking error when a neural network was used as a compensator, as compared to the case where no compensator was used.

Upon obtaining the baseline trajectory, uncertainty in the system parameters was introduced. The values of the three parameters, namely, d_1, d_2, and d_3, of the manipulator were altered, as summarized in Table 2 under the heading "Estimated". Another trial was executed with the uncertainty introduced in the manipulator parameters and without any compensating signal incorporated into the control law. Figure 23 shows the results of this trial (which is referred to as the initial trajectory) together with the baseline trajectory.

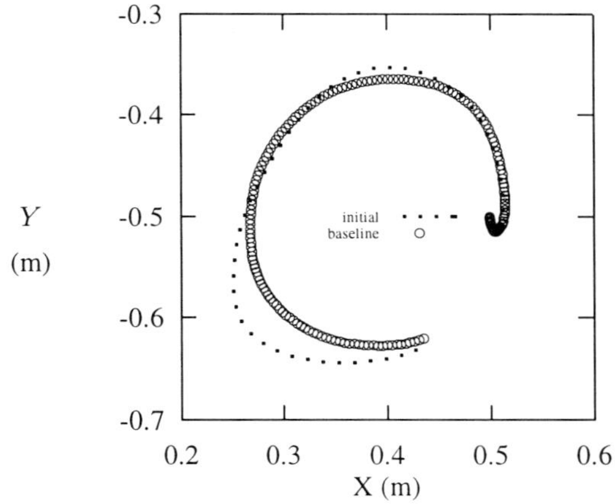

Figure 23. Baseline Trajectory and Initial Trajectory.

A series of trials was then executed with the network weights being adjusted off-line after each trial. Figures 24 shows the trajectory of the manipulator tip at the 20th trial and at the 30th trial. It can be seen that the incorporation of the neural network as an uncertainty compensator indeed improved system performance in the sense that the actual trajectory of the manipulator approached the baseline trajectory as the learning process of the neural network was iterated.

8. Discussion

8.1. Implication of Results

We have proposed an approach to improving the performance of uncertain robotic systems using neural networks. It has been shown that this approach is applicable to repetitive continuous-path robot operation. In this approach, uncertainty in the robotic system is

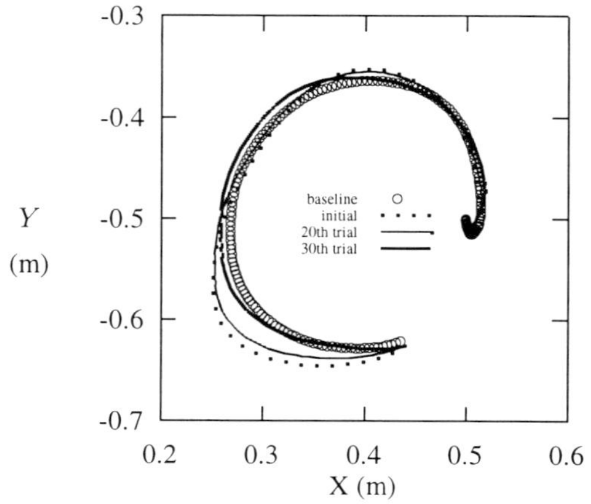

Figure 24. Task Trajectory: 20th and 30th Trial.

quantified and a neural network is used to "nullify" the uncertainty so that performance improvement can be achieved.

Using techniques from nonlinear system theory, closed-loop stability of the robotic system (incorporated with a neural network) has been analyzed. Results of the analysis confirm that the closed-loop system is stable in the sense that all signals in the system are bounded. Stability of closed-loop systems embedded with neural networks is a key issue that has not been adequately addressed in the literature. The stability analysis presented in this article offers a possible solution in resolving this issue.

A method for analyzing the performance of the robotic system (incorporated with a neural network) has been developed. Using this method, the effect of the dynamics of the neural network on the performance of the manipulator has been revealed. It has further been shown that the performance of the robotic system is improved as the learning process of the neural network is iterated.

The results of these analyses provide theoretical justification for the use of multilayer feedforward neural networks with the error-backpropagation algorithm in a feedback control system in which the dynamics of the network is coupled with that of the controlled plant. The error-backpropagation algorithm has been one of the most commonly used learning rules for neural networks. Various schemes have been proposed in the literature on the application of neural networks to robot control using this algorithm, but the important question of how the dynamics of the neural network affects the performance of the robotic system has not been addressed rigorously in the literature. The performance analysis presented in this article provides a plausible answer. The analysis confirms that, with a sufficiently small learning rate, the error-backpropagation algorithm will not destabilize the closed-loop system and will improve the performance of the robotic system as the learning process is iterated. Numerical simulations have been conducted. Results of the simulation have confirmed the conclusions of the theoretical analysis.

An experiment using a laboratory manipulator has been conducted. The results of the experiment clearly demonstrate the effectiveness of the proposed approach in improving the performance of the robotic system. The experimental implementation of the proposed approach, together with the positive experimental results, are important not only in demonstrating the effectiveness of the neural network as an uncertainty compensator, but also in demonstrating the effectiveness of a control system incorporated with a neural network for realistic robotic tasks. Many studies on the application of neural networks to system control in general and to robot control in particular rely on numerical simulation to verify the conclusions therein; very few proposed schemes have been physically implemented to verify their effectiveness. Extensive and conclusive experimental results are needed to affirm neural networks as viable engineering tools. The experiment reported in this article represents a small step in gathering such results.

8.2. Research Directions

In the context of the work presented in this article, the following directions are suggested.

Comparison with Other Approaches
An important and necessary extension of the work presented is an in-depth analysis in comparing the proposed approach with various other approaches (such as adaptive control) so that the full potential of neural networks for robot control (as discussed in Section 1) can be firmly substantiated and convincingly demonstrated.

Effect of Manipulator Link and Joint Flexibility
Throughout this work, the manipulators under consideration were treated as rigid mechanical linkages. In reality, however, link and joint flexibility exist in industrial robots. It would thus be practically meaningful to extend the proposed approach to account for the effect of link and joint flexibility of industrial robots. Such an extension will yield results that further affirm the utility of neural networks in practical robotic applications.

Reduction of the Number of Learning Trials
In analyzing the effect of neural network learning on the performance of the robotic system, i.e., Theorems 2, the condition needed to guarantee the reduction of the control error was that the difference between the weight change of two successive trials is negligible. The justification of this condition was based on the requirement that the learning rate be sufficiently small. A small learning rate, however, implies that to achieve significant performance improvement, a large number of learning trials may be required.

Since the dynamics of the network weights also depends on the gradient of the "error surface", a technique that allows variation of the learning rate based on the "steepness" of the gradient would be a potential solution to reduce the number of learning trials.

Determination of the Necessary Number of Learning Trials
The clarification of the relationship between the reduction in the control error and the reduction in the trajectory tracking error represents an important theoretical issue in under-

standing the utility of neural networks for system control in general and for robot control in particular. It is important because successful theoretical clarification of this relationship would provide a definitive answer to the question of how many trials are required in order to reduce the tracking error to within certain tolerance.

Learning Multiple Trajectories
In manufacturing operations, a robot is often required to be capable of performing more than one task. This implies that the robot controller must be able to execute more than one trajectory. The scope of the present work has been limited to the case where the robot controller is to learn to improve its performance for a single trajectory. Such a controller is clearly inadequate in a production environment. Thus, a practical extension of the present work is to develop methodologies which enable the neural-network-based controller to learn multiple trajectories and to execute these trajectories without interference.

Effective Global Convergence
Currently there is no known method that guarantees global convergence of neural network learning. This means that it is possible that the set of weights found by the error-backpropagation algorithm is optimal only locally. It is of course worthwhile to find a learning algorithm that theoretically guarantees global convergence, but it is also practically meaningful to find a way to circumvent local optimality so as to achieve effective global convergence. One plausible approach is to use multiple networks in such a way that when one network becomes "trapped" in a local optimum, another network is activated to continue the learning process. Conceptually, it can be argued that it is possible to reduce the control error to any required tolerance by using a series of networks. A method of implementing such a series of networks would significantly enhance the practicality of neural networks in robot control.

Hardware Implementation of Neural Networks for Robot Control
Currently in studies reported in the literature on neural networks for robot control, the neural networks are usually implemented in the form of software programs and executed on general-purpose digital computers. Such software implementation, however, does not take full advantage of the parallel processing capability of neural networks. Thus, in order to fully utilize the capability of neural networks for practical robotic applications, it is necessary to investigate hardware implementation of neural networks. One plausible hardware implementation of a neural network is in the form of a VLSI chip. It can be anticipated that successful synthesis of "neuro-chips" and commercial robot controllers would advance the application of industrial robots to a more sophisticated level.

Appendix A

Definitions

The following definitions are based on material presented in (Vidyasagar, 1992). The convolution of a Laplace transformable signal $f(t)$ and a transfer function matrix $M(s)$ is denoted by Mf, i.e., $Mf = (m * f)(t)$, where $*$ denotes the convolution operator. The l_2-norm of a vector $x \in R^n$, denoted by $\|x\|$, is defined as: $\|x\| = (\sum_{j=1}^{n} |x_j|^2)^{\frac{1}{2}}$. The l_2-norm of a matrix $A \in R^{n \times n}$, denoted by $\|A\|$, is defined as: $\|A\| = [\max_i \lambda_i(A^T A)]^{\frac{1}{2}}$, where λ_i denotes the eigenvalue of the matrix A. The L_∞^n-norm and the *extended* L_∞^n-space (for the truncated L_∞^n-norm) are as defined in Section 5. A system with input x and output y is said to be bounded-input bounded-output stable (or BIBO stable) if for every $x \in L_\infty$, $y \in L_\infty$. The L_∞^n-norm of a transfer matrix P is defined as: $\|P\|_\infty = sup_{x \in L_\infty^n - 0} \frac{\|Px\|_\infty}{\|x\|_\infty}$. Let β denote the norm $\|P\|_\infty$, then $\|Px\|_{T\infty} \le \beta \|x\|_{T\infty}$.

Appendix B

Proof of Theorem 1

The method presented in (Spong & Vidyasagar, 1987) is utilized in constructing this proof. Through algebraic operations, the system (10) can alternatively be expressed as

$$\dot{\bar{e}} = \bar{A}\bar{e} + B(\bar{\eta} + u) \tag{B.1}$$

where $\bar{e} = \begin{bmatrix} q - q^d \\ \dot{q} - \dot{q}^d \end{bmatrix}$, $\bar{A} = \begin{bmatrix} 0 & I \\ 0 & 0 \end{bmatrix}$, $B = \begin{bmatrix} 0 \\ I \end{bmatrix}$, $\bar{\eta} = \eta_0 + Eu$, $\eta_0 = E\ddot{q}^d + M^{-1}\Delta h$, $u = K\bar{e} + v$, $E = M^{-1}\hat{M} - I$, $\Delta h = \hat{h} - h$, $K = [-K_p, -K_v]$, and v is the neural network output.

We make the following assumptions (Spong & Vidyasagar, 1987) regarding the nominal model of the manipulator. *(A1)*: For the inertia matrix M of the manipulator, there exist constants M_1 and M_2 such that $M_1 \le \|M^{-1}\|_{T\infty} \le M_2 < \infty$. *(A2)*: There exists a nonnegative constant $\alpha < 1$ such that $\|M^{-1}\hat{M} - I\|_{T\infty} \le \alpha$. *(A3)*: There exist nonnegative constants δ and ρ such that $\|\hat{h} - h\|_{T\infty} \le \delta\|x\|_{T\infty} + \rho$, where $x = [q^T, \dot{q}^T]^T$.

Let $G(s) = (sI - \bar{A})^{-1}B$. Then (B.1) can be expressed as: $\bar{e} = G\epsilon$, with $\epsilon = \bar{\eta} + u$, $u = K\bar{e} + v$, and $\bar{\eta} = \eta_0 + Eu$. Through algebraic operations, we obtain $\bar{e} = (I - GK)^{-1}G\bar{\eta} + (I - GK)^{-1}Gv$, $u = K(I - GK)^{-1}G\bar{\eta} + (K(I - GK)^{-1}G + I)v$. Let $P_1 = (I - GK)^{-1}G$, $P_2 = K(I - GK)^{-1}G$, and $P_3 = K(I - GK)^{-1} + I$, then we can express $\bar{e}$ and u as: $\bar{e} = P_1\bar{\eta} + P_1 v$, and $u = P_2\bar{\eta} + P_3 v$. Now taking the truncated L_∞-norm yields

$$\|\bar{e}\|_{T\infty} \le \beta_1\|\bar{\eta}\|_{T\infty} + \beta_1\|v\|_{T\infty}, \tag{B.2}$$

$$\|u\|_{T\infty} \le \beta_2\|\bar{\eta}\|_{T\infty} + \beta_3\|v\|_{T\infty}, \tag{B.3}$$

where $\beta_1 = \|(I - GK)^{-1}G\|_{T\infty}$, $\beta_2 = \|K(I - GK)^{-1}G\|_{T\infty}$, and $\beta_3 = \|K(I - GK)^{-1}G + I\|_{T\infty}$. Since $\eta_0 = E\ddot{q}^d + M^{-1}\Delta h$, from the modeling assumptions, we have $\|\eta_0\|_{T\infty} \le M_2\delta\|\bar{e}\|_{T\infty} + b$, where $b = \alpha\|\ddot{q}^d\|_{T\infty} + M_2\delta\|x^d\|_{T\infty} + M_2\rho$, and $x^d = [(q^d)^T, (\dot{q}^d)^T]^T$. It follows that $\|\bar{\eta}\|_{T\infty} \le M_2\delta\|\bar{e}\|_{T\infty} + \alpha\|u\|_{T\infty} + b$. Since the output of the neural network is bounded by construction (due to the $tanh$ activation function of the output units), let $\phi = \|v\|_{T\infty}$. Hence $\begin{bmatrix} \|\bar{\eta}\|_{T\infty} \\ \|u\|_{T\infty} \end{bmatrix} \le \begin{bmatrix} M_2\delta\beta_1 & \alpha \\ \beta_2 & 0 \end{bmatrix} \begin{bmatrix} \|\eta_2\|_{T\infty} \\ \|u\|_{T\infty} \end{bmatrix} +$

$\begin{bmatrix} M_2\delta\beta_1\phi + b \\ \beta_3\phi \end{bmatrix}$. Let $Q = \begin{bmatrix} M_2\delta\beta_1 & \alpha \\ \beta_2 & 0 \end{bmatrix}$. Then $\det(I - Q) = 1 - M_2\delta\beta_1 - \alpha\beta_2 \equiv \Delta_m$. If $\Delta_m > 0$, then

$$\begin{bmatrix} \|\bar{\eta}\|_{T\infty} \\ \|u\|_{T\infty} \end{bmatrix} \leq \frac{1}{\Delta_m} \begin{bmatrix} 1 & \alpha \\ \beta_2 & 1 - M_2\delta\beta_1 \end{bmatrix} \begin{bmatrix} M_2\delta\beta_1\phi + b \\ \beta_3\phi \end{bmatrix}. \tag{B.4}$$

From (B.2) and (B.4), we obtain $\|\bar{e}\|_{T\infty} \leq \frac{\beta_1 b}{\Delta_m} + \frac{\beta_1 \phi}{\Delta_m}(\beta_1 M_2\delta + \alpha\beta_3)$. Therefore, if the condition $\Delta_m > 0$ is satisfied, then $\bar{\eta}$, u, and $\bar{e}$ are bounded. Now recall that $\beta_1 = \|(I - GK)^{-1}G\|_{T\infty}$, and $\beta_2 = \|K(I - GK)^{-1}G\|_{T\infty}$. It can be seen that the condition $\Delta_m > 0$ can be satisfied by selecting sufficiently large values for K such that β_1 approaches zero and β_2 approaches unity simultaneously.

Appendix C

Proof of Corollary 1

From (1) and (5), we have $M\ddot{q}+h = \hat{M}(\ddot{q}^d+u)+\hat{h}$, where $u = K_v(\dot{q}^d-\dot{q})+K_p(q^d-q)+v$. So $\ddot{q} = E(\ddot{q}^d + u) + \ddot{q}^d + u + M^{-1}\Delta h$, where E and Δh are as defined in (B.1). From the modeling assumptions $A.1$-$A.3$, we have

$$\begin{aligned} \|\ddot{q}\|_{T\infty} &\leq \|E\|_{T\infty}\|\ddot{q}^d + u\|_{T\infty} + \|\ddot{q}^d + u\|_{T\infty} + \|M^{-1}\|_{T\infty}\|\Delta h\|_{T\infty} \\ &\leq (1+\alpha)(\|\ddot{q}^d\|_{T\infty} + \|u\|_{T\infty}) + M_2\delta\|\bar{e}\|_{T\infty} + M_2\delta\|x^d\|_{T\infty} + M_2\rho, \end{aligned}$$

where $x^d = [(q^d)^T, (\dot{q}^d)^T]^T$. Since u and $\bar{e}$ have been proved to be bounded (Theorem 1), it can be concluded that $\ddot{q}$ is bounded.

Appendix D

Proof of Corollary 2

From (9), we have $\Delta v = \ddot{q}^d - \ddot{q} + K\bar{e}$, where K is as specified in (B.1). Therefore, $\|\Delta v\|_{T\infty} \leq \|\ddot{q}^d\|_{T\infty} + \|\ddot{q}\|_{T\infty} + \|K\bar{e}\|_{T\infty}$. Since $\ddot{q}$ and $\bar{e}$ have been proven to be bounded, it can be concluded that Δv is also bounded.

Appendix E

Proof of Corollary 3

As described in Section 4 (with reference to Figure 4), we have

$$\dot{W}_{ij} = \lambda_n \Delta v_i g'(v_i)\bar{v}_j, \dot{R}_{jk} = \lambda_n g'(\bar{v}_j) \sum_{i=1}^{I_n} (\Delta v_i g'(v_i) W_{ij}),$$

and $\dot{S}_{kl} = \lambda_n g'(\bar{\bar{v}}_k) \sum_{j=1}^{J_n} \left(g'(\bar{v}_j) \sum_{i=1}^{I_n} (\Delta v_i g'(v_i) W_{ij} R_{jk})\right)$, where $g'(\cdot) = \frac{\partial g(\cdot)}{\partial(\cdot)}$. Since $\dot{W}_{ij}$ depends on $\lambda_n, \Delta v_i, g'$, and $\bar{v}_j$, which are all bounded, it can be concluded that $\dot{W}_{ij}$ and consequently W_{ij} are bounded during a given trial. Similarly, it can be concluded that $\dot{R}_{jk}, R_{jk}, \dot{S}_{kl}$ and S_{kl} are also bounded during a given trial.

Appendix F

Proof of Proposition 1

From Section 4, the dynamics of the neural network can be expressed as: $\dot{W}_{ij} = \lambda_n \Delta v_i g'(v_i)\bar{v}_j, \dot{R}_{jk} = \lambda_n g'(\bar{v}_j)\sum_{i=1}^{I_n}(\Delta v_i g'(v_i)W_{ij})$, and $\dot{S}_{kl} = \lambda_n g'(\bar{\bar{v}}_k)\sum_{j=1}^{J_n}\left(g'(\bar{v}_j)\sum_{i=1}^{I_n}(\Delta v_i g'(v_i)W_{ij}R_{jk})\right)$. During the subsequent development, the symbols λ_w, λ_r, and λ_s are used (instead of the general notation λ_n) to indicate that these learning rates are specifically associated with W_{ij}, R_{jk}, and S_{kl} respectively. Since during a given trial p, the signals driving the weights are bounded, i.e., $\|\Delta v_i(p,\xi)\|_{T\infty} \leq \gamma_p < \infty$, $\|v_i(p,\xi)\|_{T\infty} \leq \tilde{c}$, $\|\bar{v}_j(p,\xi)\|_{T\infty} \leq \tilde{c}$, $\|\bar{\bar{v}}_k(p,\xi)\|_{T\infty} \leq \tilde{c}$, and $\|g'(\cdot)\|_{T\infty} \leq \tilde{c}$, where $\tilde{c}$ is the largest scaling factor, hence $\|\dot{W}_{ij}(p,\xi)\|_{T\infty} = \|\lambda_w \Delta v_i(p,\xi)g'(v_i(p,\xi))\bar{v}_j(p,\xi)\|_{T\infty} \leq \lambda_w \tilde{c}^2\gamma_p$, and $\|W_{ij}(p,\xi)\|_{T\infty} = \left\|W_{ij}(p,0) + \int_0^\xi \dot{W}_{ij}(p,\sigma)d\sigma\right\|_{T\infty} \leq \lambda_w\tilde{c}^2\gamma_p\xi + \gamma_p^w \equiv \psi_p^w$. Similarly, $\|\dot{R}_{jk}(p,\xi)\|_{T\infty} \leq \lambda_r\tilde{c}^2 I_n\gamma_p\psi_p^w$, $\|R_{jk}(p,\xi)\|_{T\infty} \leq \lambda_r\tilde{c}^2 I_n\gamma_p\psi_p^w\xi + \gamma_p^r \equiv \psi_p^r$, and $\|\dot{S}_{kl}(p,\xi)\|_{T\infty} \leq \lambda_s\tilde{c}^3 I_n J_n\gamma_p\psi_p^w\psi_p^r$. From the definitions presented in Section 5, we have $\Delta w_{ij}(p,\xi) = \int_0^\xi \dot{W}_{ij}(p,\sigma)d\sigma$, $\Delta r_{jk}(p,\xi) = \int_0^\xi \dot{R}_{jk}(p,\sigma)d\sigma$, and $\Delta s_{kl}(p,\xi) = \int_0^\xi \dot{S}_{kl}(p,\sigma)d\sigma$.

Now for two successive trials p and $(p+1)$, $\|\Delta w_{ij}(p+1,\xi) - \Delta w_{ij}(p,\xi)\|_{T\infty} \leq \left\|\int_0^\xi \dot{W}_{ij}(p+1,\sigma)d\sigma\right\|_{T\infty} + \left\|\int_0^\xi \dot{W}_{ij}(p,\sigma)d\sigma\right\|_{T\infty} \leq \lambda_w\tilde{c}^2(\gamma_{p+1}+\gamma_p)$. Therefore, to satisfy $\|\Delta w_{ij}(p+1,\xi) - \Delta w_{ij}(p,\xi)\|_{T\infty} << 1$, λ_w can be chosen such that $\lambda_w << \frac{1}{\tilde{c}^2(\gamma_{p+1}+\gamma_p)}$. Similarly, λ_r and λ_s can be chosen as: $\lambda_r << \frac{1}{\tilde{c}^2 I_n(\gamma_{p+1}\psi_{p+1}^w + \gamma_p\psi_p^w)}$, and $\lambda_s << \frac{1}{\tilde{c}^3 I_n J_n(\gamma_{p+1}\psi_{p+1}^w\psi_{p+1}^r + \gamma_p\psi_p^w\psi_p^r)}$. It can be seen that by choosing λ_n to be $\lambda_n = \min[\lambda_w, \lambda_r, \lambda_s]$, the general condition, $\|\Delta\theta_m(p+1,\xi) - \Delta\theta_m(p,\xi)\|_{T\infty} << 1$, can be met.

Acknowledgments

This work is financially supported by the Institute of Robotics and Intelligent Systems, the Ontario Graduate Scholarship, and the Natural Sciences and Engineering Research Council of Canada. The authors wish to thank Professor B. Benhabib for the use of the computing facility at the Computer Integrated Manufacturing Laboratory, Department of Mechanical Engineering, University of Toronto.

References

Arai, F., Rong, L. & Fukuda, T. (1993). Asymptotic Convergence of Feedback Error Learning Method and Improvement of Learning Speed. *Proceedings of the 1993 IEEE/RSJ International Conference on Intelligent Robots and Systems*, Yokohama, Japan, 761-767.

Cılız, M.K.& Işık, C. (1990). Trajectory Following Control of Robotic Manipulators Using Neural Networks. *Proceedings of the 5th IEEE International Symposium on Intelligent Control*, Philadelphia, Pennsylvania, 536-540.

Corless, M.& Leitmann, G. (1981). Continuous State Feedback Guaranteeing Uniform Ultimated Boundedness for Uncertain Dynamic Systems. *IEEE Transactions on Automatic Control*, AC-26, No. 5, 1139-1144.

Craig, J. (1986). *Introduction to Robotics: Mechanics and Control*. Reading, MA: Addison-Wesley.

Cybenko, G. (1989). Approximation by Superposition of a Sigmoidal Function. *Mathematics of Control, Signals, and Systems*, Vol. 2, 303-314.

Denker, J., Schwartz, D., Witter, B., Solla, S., Howard, R. & Jackel, L. (1987). Large Automatic Learning, Rule Extraction, and Generalization. *Complex Systems*, Vol. 1, 877-922.

Dorf, R.C. (1992). *Modern Control Systems*. Reading, Massachusetts: Addison-Wesley Publishing Company.

Glesner, M.& Pöchmüller, W. (1994). *Neurocomputers: An Overview of Neural Networks in VLSI*. London: Chapman & Hall.

Gomi, H.& Kawato, M. (1990). Learning Control for a Closed Loop System Using Feedback-Error-Learning. *Proceedings of the 29th Conference on Decision and Control*, Honolulu, Hawaii, 3289-3294, 1990.

Kuan, A.& Bavarian, B. (1992). Control of Universal 3DOF Robot Manipulator: A Neural Network Approach. *Proceedings of the Japan-USA Symposium on Flexible Automation*, San Francisco, CA, 249-255.

Lawryshyn, Y.A. (1993). *Robot Link and Contact Surface Dynamic Interaction: An Experimental Study of Contact Stability*. M.A.Sc. Thesis, Department of Mechanical Engineering, University of Toronto.

Lokhorst, D. M. (1990). *An Approach to Force and Position Control of Robots: Theory and Experiments*. M.A.Sc. Thesis, Department of Mechanical Engineering, University of Toronto, Toronto, Canada.

McClelland, J.L.& Rumelhart, D.E. (Eds.), (1986). *Parallel Distributed Processing*. Vol. 1, Cambridge, Massachusetts: MIT Press.

Miller, W.T., Sutton, R.S. & Werbos, P.J. (Eds.), (1990). *Neural Networks for Control*. Cambridge, Massachusetts: MIT Press.

Muller, B. & Reinhardt, J. (1990). *Neural Networks: An Introduction*. Berlin: Springer-Verlag, 1990.

Okuma, S.& Ishiguro, A. (1990). A Neural Network Compensator for Uncertainties of Robotic Manipulators. *Proceedings of the 29th Conference on Decision and Control*, Honolulu, Hawaii, 3303-3307.

Ortega, R.& Spong, M.W. (1989). Adaptive Motion Control of Rigid Robots: a Tutorial. *Automatica*, Vol.25, No.6, 877-888.

Spong, M.W.& Vidyasagar, M. (1989). *Robot Dynamics and Control*. New York: John Wiley & Sons.

Spong, M.W.& Vidyasagar, M. (1987). Robust Linear Compensator Design for Nonlinear Robotic Control. *IEEE Journal of Robotics and Automation*, Vol. RA-3, No. 4, 345-351.

Vidyasagar, M., (1992). *Nonlinear Systems Analysis*. 2nd ed., New Jersey: Prentice Hall.

Yabuta, T.& Yamada, T. (1991). Learning Control Using Neural Networks. *Proceedings of the 1991 IEEE International Conference on Robotics and Automation*, Sacramento, California, 740-745.

Yanovski, G., (1991). *Experimental Analysis of a Robust Centralized/Decentralized Controller for Robotic Manipulators*. M.E. Thesis, Department of Electrical Engineering, University of Toronto.

Zomaya, A.Y.& Nahban, T.M. (1993). Centralized and Decentralized Neuro-Adaptive Robot Controllers. *Neural Networks*, Vol. 6, 223-244.

Machine Learning, 23, 221–249 (1996)

Learning Controllers for Industrial Robots

C. BAROGLIO baroglio@di.unito.it
Dipartimento di Informatica, Università di Torino, C.so Svizzera 185, 10149 Torino, Italy

A. GIORDANA attilio@di.unito.it
Dipartimento di Informatica, Università di Torino, C.so Svizzera 185, 10149 Torino, Italy

M. KAISER kaiser@ira.uka.de
University of Karlsruhe, Institute for Real-Time Computer Systems & Robotics

M. NUTTIN nuttin@mech.kuleuven.ac.be
Katholieke Universiteit Leuven, Department of Mechanical Engineering, Division PMA

R. PIOLA piola@di.unito.it
Dipartimento di Informatica, Università di Torino, C.so Svizzera 185, 10149 Torino, Italy

Editors: Judy A. Franklin, Tom M. Mitchell, and Sebastian Thrun

Abstract. One of the most significant cost factors in robotics applications is the design and development of real-time robot control software. Control theory helps when linear controllers have to be developed, but it doesn't sufficiently support the generation of non-linear controllers, although in many cases (such as in compliance control), nonlinear control is essential for achieving high performance. This paper discusses how Machine Learning has been applied to the design of (non-)linear controllers. Several alternative function approximators, including Multilayer Perceptrons (MLP), Radial Basis Function Networks (RBFNs), and Fuzzy Controllers are analyzed and compared, leading to the definition of two major families: Open Field Function Approximators and Locally Receptive Field Function Approximators. It is shown that RBFNs and Fuzzy Controllers bear strong similarities, and that both have a symbolic interpretation. This characteristic allows for applying both symbolic and statistic learning algorithms to synthesize the network layout from a set of examples and, possibly, some background knowledge. Three integrated learning algorithms, two of which are original, are described and evaluated on experimental test cases. The first test case is provided by a robot KUKA IR-361 engaged into the "peg-into-hole" task, whereas the second is represented by a classical prediction task on the Mackey-Glass time series. From the experimental comparison, it appears that both Fuzzy Controllers and RBFNs synthesised from examples are excellent approximators, and that, in practice, they can be even more accurate than MLPs.

Keywords: Robotics, Neural Networks, Fuzzy Controllers, Multistrategy Learning

1. Introduction

Traditionally, robotics has been a great challenge for artificial intelligence and machine learning, since it offers the opportunity to design truly intelligent machines that interact with their environment and exhibit human-like capabilities. Nowadays, the increasing availability of sophisticated sensor systems for both manipulation and mobile robots makes the development of smarter and more adaptive robots also technically possible. Such high-performance systems are opening new markets to the robot technology, e. g. the recently emerging market for intelligent service systems. However, the cost of programming such robots is high, and learning capabilities, i.e. the ability of the robot to gain profit from its experiences, become indispensable for fully exploiting the robot's potential autonomy (Kaiser et al., 1995b).

In this paper, we describe the results of an investigation done in the framework of the Esprit project No. 7274, B-LEARN II. The fundamental goal of the project is the enhancement of current industrial robots by incorporating learning capabilities on all levels of robot control. The expected benefits are a decrease in the cost for developing the control software itself and an increase in robots' reliability and flexibility. B-LEARN II is a wide-spectrum project, structured into several subprojects related to different applicative fields (Kaiser et al., 1994). In the research presented here, we focus on controllers for robots employed in assembly lines, performing tasks (such as cutting and deburring), in which the same operation is repeated thousands of times, and information coming from rough sensors only, such as force and torque sensors, is exploited. In this kind of application, it is not interesting to optimize any problem solving capability (as one might think reading about the application of "AI techniques"), but to achieve high accuracy as well as high speed in accomplishing simple operations such as following a contour.

In the traditional approach, such robots are usually controlled by means of linear controllers. However, the simplifications involved in linear controller design are not always justified. Especially the physical characteristics of the robot and the existing hardware and software limitations that must be taken into account in the real world introduce nonlinearities that are difficult to model.

The aim of this work is to assess the usability of ML techniques to increase the robot's performance by means of nonlinear control, to ease the development of these controllers, and to decrease the cost for realizing a new application. Learning to control robots on a low level means learning continuous, real-valued functions: it is a regression problem. In the literature, regression has been tackled by means of different techniques such as neural networks (Rumelhart & McClelland, 1986; Rumelhart et al., 1985), prediction trees, statistics (Cramer, 1974; Quinlan, 1993), regression trees (Breiman et al., 1984), and fuzzy logics (Berenji, 1992); moreover, it has been seen either as a supervised or as a reinforcement learning task (Barto et al., 1983; Gullapalli, 1990).

The aim of this paper is to perform a comparative analysis of the different methods in order to find out both the function approximators and the learning algorithms that are suitable for the class of robotic applications described above. The major result of this research is the characterization of a class of approximators (including Radial Basis Function Networks (Poggio & Girosi, 1990) and Fuzzy Controllers (Berenji, 1992)), that we will call Locally Receptive Field Networks (LRFN), which exhibits properties that are more interesting than those of the well-known multilayer perceptron. More specifically, LRFNs, beyond being excellent approximators, can be interpreted in terms of symbolic knowledge and can be synthesised by means of standard symbolic and/or statistic learning algorithms. Furthermore, LRFNs allow for incremental learning, which is an important issue especially when realizing adaptive controllers (Miller et al., 1990).

An important novelty is the multistrategy method we propose for learning LRFNs from examples of correct behavior and background knowledge. The emphasis will be given to supervised learning, which can also be a basis for reinforcement learning architectures, as shown by experiments based on Williams REINFORCE algorithms (Williams, 1992; Nuttin et al., 1995).

The paper is organized as follows. The next section investigates the requirements posed by the robotic applications we are interested in. Section 3 performs a comparative analysis of the main methods available for regression, and tries to set a unifying view. Section 4 describes the supervised learning methods, and Section 5 presents an experimental evaluation of both the approximators and the corresponding learning techniques. Finally, the results are discussed in Section 6 and the conclusions are taken in Section 7.

2. Learning Compliance: a Key Issue in Assembly Robots

In this section we will investigate the nature of the control activity in a robot oriented to industrial production in an assembly line. As it has been introduced above, we assume them to be devoted to repetitive tasks, guided by rough sensors[1].

Typically, the program controlling a robot employed in an industrial assembly line can be considered as organized in two layers. The highest one establishes the basic cycle executed by the robot. This layer can be realized by a traditional robot program, but it can as well be given as a *dispatcher* acting as a plan execution unit. The lowest level closes the loop between sensing and action and takes care of the compliant behavior of the robot. To this aim, data originating from force and torque sensors are employed.

In general, *compliant motion* refers to tasks that require the robot to establish or maintain contact with its environment such that the capability of continuously adapting the robot's action to its perception is needed (Mason, 1981). As an example, we can think of executing the insertion of a peg into a hole. The quality of this action can be evaluated as a trade-off between the global time required for achieving the goal state *peg-inserted*, and the strength of the forces which should remain as low as possible during the whole insertion. The solutions for dealing with this problem in the presence of uncertainty are the use of intelligently designed passive compliance by means of elastic joints as well the realization of an active compliance mechanism, for instance by means of conventional PID-controllers[2].

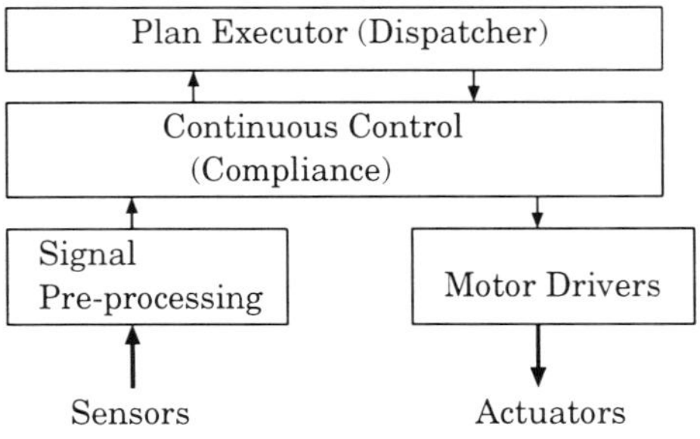

Figure 1. Two layer control architecture for an assembly robot.

Such a controller tries to keep the system in a nominal state; if the state changes, the controller reacts with a signal proportional to the difference between the nominal state and the actual one. In our case, for instance, a linear controller could be used to keep the contact force between the surfaces constant. Unfortunately, a reaction proportional to the error is not always the best solution to restore the nominal state as quickly as possible. For the peg-into-hole task, it has even been proven that the optimal controller must be non linear (Asada, 1990).

However, while control theory offers a good basis for developing linear controllers, it does not support the design of non-linear controllers sufficiently. Therefore, empirical tools such as Fuzzy Controllers (Berenji, 1992), that allow non-linear control functions to be realized while avoiding the burden of developing a mathematical model, are attracting a lot of attention. In this context, we claim that ML can be of substantial help for both the reduction of the development costs of an empirical solution as well as for the improvement of the implementation quality.

When and how can ML techniques be applied to this aim? A possible answer can be obtained by examining the knowledge sources available for an empirical approach to control synthesis. A first knowledge source can be a qualitative theory about the control problem. In fact, although it is difficult, or even impossible to derive an analytical control function from a quantitative theory, it is easy to find some expert having a qualitative idea of its general shape. In general, this knowledge is the starting point for manual fuzzy controller synthesis (Bonissone & Chiang, 1993; Nuttin et al., 1994).

A second source is represented by the ability of a human operator (or of some other machine) of executing the control task him/herself. It is extremely hard to explicitly formalize this knowledge as a theory; nevertheless, it is easy to apply it to generate examples of control, and even from bad examples, control knowledge can be obtained (Kaiser et al., 1995a).

Considering this second knowledge source, we immediately recognize the typical supervised learning framework, that can be applied for regression as well as for classification problems. We would like, however, to stress the importance of exploiting qualitative theories too. As a matter of fact, in many cases, it is difficult to have a set of examples of control actions covering all possible cases (a necessary condition for learning a complete control function). In this case, using a domain theory can help to reduce the amount of training examples significantly. The success of the manual synthesis (Bonissone & Chiang, 1993; Yih & Shieh, 1992) is an implicit proof of this claim.

However, learning only from examples provided by a teacher has an intrinsic limitation: it allows the synthesis of controllers performing *at most* as well as the teacher him/herself. Actually, taking an optimistical view (filtering the noise from the examples and using a good domain theory), we might even achieve a performance better than that of the teacher, but this does not mean to overcome the inherent limitation. The only way to get rid of this problem is to widen the learning framework by means of an integrated approach (Kaiser & Kreuziger, 1994), using also other learning paradigms (such as reinforcement learning) that are capable of continuously improving the controller's performance. In this respect, an important feature of the learning algorithms and the representations that are to be used is their support of *incremental* learning, to allow an expert to supply his/her knowledge progressively as well as to allow for on-line refinement of the control function.

In the following sections, we will analyse and compare several tools for approximating continuous functions. We will investigate which of them meet the requirements mentioned above and are therefore most suitable. We'll also identify those that allow for incremental learning and can use domain knowledge and examples. For two major reasons, the emphasis will be on supervised learning algorithms. Firstly, many reinforcement learning techniques do actually rely on supervised learning algorithms, such as backpropagation. Secondly, supervised learning is a good means for initializing a robot's knowledge off-line. The robots we are dealing with are delicate and potentially harmful. Therefore, we would like a robot controller to be at least partially operational from the very beginning. A reinforcement learning approach based on random exploration without any knowledge about the environment is simply not feasible: the robot will damage itself (or it will cause too many troubles) before learning anything. Therefore, the only viable solution is to off-line generate a controller that already fulfills minimal operationality requirements before beginning to learn by reinforcement.

3. Approximating Control Functions

The task of learning to approximate continuous functions has been investigated in many fields, using alternative approaches such as statistics (Breiman et al., 1984; Specht, 1988; Specht, 1990), connectionism (Rumelhart & McClelland, 1986; Barto et al., 1990), fuzzy logics (Zadeh, 1992; Berenji, 1992), and symbolic machine learning (Sammut et al., 1992).

In applicative domains such as robotics and automated controls, two approximators attracted a great attention: Multi-Layer Perceptrons (MLPs) and Fuzzy Controllers (FCs). The reason for their popularity is mostly due to their robustness and inherent simplicity, which allows applications to be quickly developed without requiring specific skills from the user. Furthermore, low cost hardware supporting real time computation is available for both architecture types.

Even if MLPs and/or FCs seem to be an obvious choice in a project with an industrial perspective, we are interested in examining the problem of control function approximation from a more general viewpoint. Therefore, many different architectures have been analysed and compared. Apart from regression and prediction trees, all existing approximators (independent on the approach they originate from) can be represented by some multi-layered network, in which nodes (or neurons) computing a simple function (activation function) are connected by means of weighted links. The differences among the various kinds of approximators reside in the activation functions and in the methods used to build the network and to update the weights (the learning algorithms).

The activation functions can be grouped into two major families: *Open Field Functions* (OFFs) and *Locally Receptive Field Functions* (LRFFs). The OFFs exhibit an anti-symmetry property with respect to a hyperplane splitting the input domain into two semi-spaces, whereas LRFFs have radial symmetry. A good example of the first family is the well-known sigmoid function $\sigma(x) = \frac{1}{1+e^{-x}}$, widely used in MLPs and in all their descendants. Examples of the second family are the multidimensional Gaussian functions used in Radial Basis Function Networks (RBFNs) (Poggio & Girosi, 1990) or in Statistical Neural Networks (SNN) (Specht, 1988; Specht, 1990). The choice of an activation function belonging to one or the other family entails properties fundamentally different.

As it has been proven in (Hornik et al., 1989), a universal function approximator can be constructed using only three layers of nodes (input, hidden and output), if the activation function in the hidden layer is non-linear. In Figure 2 we compare the structure of two networks of this kind: one based on sigmoidal and the other on bell-shaped functions. From a topological viewpoint, they look identical. Nevertheless, the way they encode the target function is very different.

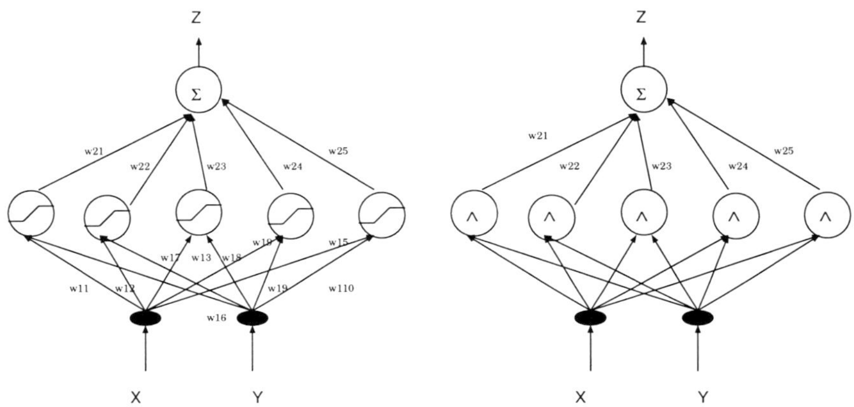

(a) (b)

Figure 2. Multilayer perceptron (a) and Radial Basis Function Network (b) architecture. The fundamental difference is in the activation function in the hidden layer neurons. In both cases the output neuron performs the sum $\sum_{i=1}^{N} w_i o_i$.

Considering the activation function $O_i = \sigma(\sum_{j=1}^{N} w_{ij} I_j)$ of the MLP's hidden units, we see that each neuron (apart from a narrow region where the sigmoid transient occurs) splits the input space into two semi-spaces where its output is significantly close to 1 or to 0. Only the semispace where the output is close to 1 contributes to the value of the target function.

In case of the LRFN, each neuron is associated to a closed region in the input space (hyper-ellipse), where its response is significantly greater than zero, and dominates over that of every other node. The contribution to the approximation of the target function comes essentially from this region.

In order to understand the different behaviors of the two network types, suppose to modify a weight between two nodes in the MLP. The effect affects an *infinite* region of the input space and can also affect a large part of the co-domain of the target function. On the contrary, changing the amplitude or shifting the position of the activation function in a LRFN will have an effect *local* to the corresponding region. This *locality property* of the LRFN allows for the network layout to be incrementally constructed (e.g. (del R. Millán, 1994)) by adjusting existing neurons and/or adding new ones. As every change has a local effect, the knowledge encoded in the other parts of the network is not lost; so, it will not be necessary to go through a global revision process.

Another property of LRFNs that is related to locality is the tendency to *undergeneralize*. Depending on the way the hidden layer (the layout) is constructed, it is possible to have large regions of the input space that are not covered by any neuron. If the current input situation falls in such regions, the output of the network will constantly remain close to the null value (usually zero), because no neuron will fire. On the contrary, every neuron of a MLP covers an infinite region of the input space, so the network will usually be active in any situation. A trained MLP has, therefore, the tendency to overgeneralize. Depending on the application (e.g., if it is safety-critical or not), under(over)generalization can be either an advantage or a disadvantage.

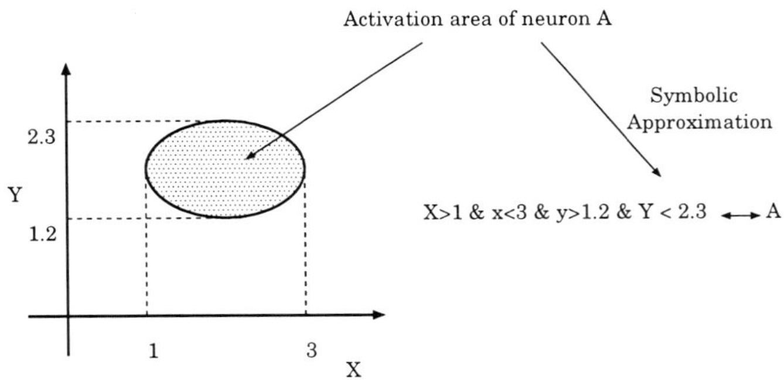

Figure 3. The closed region where the LRF-Function is dominant can be roughly approximated by a conjunctive logical assertion.

Finally, an important property of LRFNs (as opposed to OFFNs, which, with few exceptions (Towell et al., 1990), are black-boxes w.r.t. their interpretation) is the possibility to directly obtain a symbolic interpretation of the hidden neurons. In fact, a closed region in the input space can be approximated using a hyper-rectangle which, in turn, can be described by a propositional formula, as shown in Figure 3. This property has already been exploited for *manually* encoding fuzzy controller's layouts using qualitative domain knowledge (Bonissone & Chiang, 1993). We will show how symbolic learning algorithms can be used to *automatically* build excellent layouts for LRFNs.

In the following, four specific network architectures will be introduced: Time-Delay Neural Networks, Radial Basis Function Networks, Time-Delay Radial Basis Function Networks, and a Fuzzy Controller architecture belonging to the LRFN family. It is worth noticing that the choice of these four types originates from an earlier, much wider, explorative investigation.

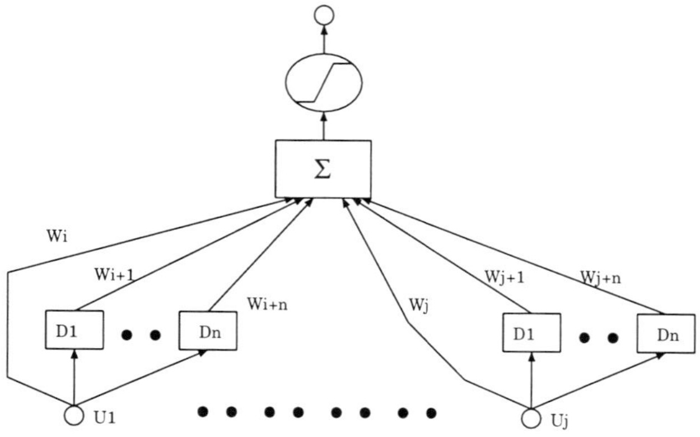

Figure 4. Basic structure of a TDNN elementary unit.

3.1. Time-Delay Neural Networks

The Time-Delay Neural Networks (TDNNs) have been introduced by A. Waibel (Waibel et al., 1989) to handle time variant signals, and are basically an extension of the multilayer perceptron. When coping with time variant signals, it is important to consider the short term history in order to predict the future values. The idea followed in TDNNs is based on the invention of the *time delays,* i.e., of a *First-In First-Out* memory, that enable a neuron to store the history of its input signals (Fig. 4).

Each delayed input is associated to a weight that can be trained using the generalized delta rule in combination with the backpropagation algorithm. Thus a neuron can detect relationships between the current and the former input values, i.e., the the network can adapt to *spatio-temporal patterns.* Delaying neurons' input in a hidden or output layer is similar to multiplying the layers in a multilayer perceptron. This may help in dealing with pattern scaling and translation. Of course, delaying also results in a higher network complexity.

3.2. Radial Basis Function Networks

RBFNs (Fig. 2b) are a family of LRF-Networks architecture that has already been investigated by several authors and recently received a strong mathematical foundation thanks to Poggio and Girosi (Poggio & Girosi, 1990).

Principally, a RBFN consists of an input layer holding the current input values, a hidden layer representing a number of closed regions (often called *kernels*) in the input space, and an output layer integrating the signals calculated by the hidden layer units. This calculation is performed on the basis of two parameters specific of each hidden unit: center and width.

The hidden neuron associated to a particular area returns the highest output value if the inputs it receives are close to its center. The actual output of a hidden neuron is determined by its activation function (mostly a multidimensional Gaussian function or a function representing a hypersphere or a hypercube), applied to a distance measure, usually computed by means of the Euclidean rule, i.e., for any input vector $\vec{x}$, a RBFN computes an output y, by means of equation

$$y = f(\vec{x}) = s((\sum_{i=1}^{n} w_i r_i(\vec{x})) + w_0)$$

where s denotes the activation function of the output neuron (which is often the identity), the w_i's are the weights on the connections between the hidden layer and the output neuron, w_0 is a bias and $r_i(\vec{x})$ is the activation function for hidden neuron i. Usually,

$$r_i(\vec{x}) = e^{-(\frac{||\vec{x}-\vec{\mu_i}||}{\sigma_i})^2}.$$

Here, $\vec{\mu_i}$ is the center and σ_i the width of neuron i, the transfer function r_i is a multidimensional Gaussian. Alternatively, a multidimensional Gaussian can be replaced by the product of *N* unidimensional Gaussians

$$r_i(\vec{x}) = e^{-(\sum_{j=1}^{dim(\vec{x})}(\frac{x_j-\mu_{ij}}{\sigma_{ij}})^2)},$$

where $dim(\vec{x})$ is the dimension of the input space. It has been proven that these networks are capable of universal approximation on $\Re$ or on compact subsets of $\Re$ (Park & Sandberg, 1993), given suitable centers μ_i.

3.3. *Time-Delays in Radial Basis Function Networks*

For handling time-variant inputs, *time delays* can also be introduced in RBFNs. Such an extension has been proposed by Berthold (Berthold, 1994).

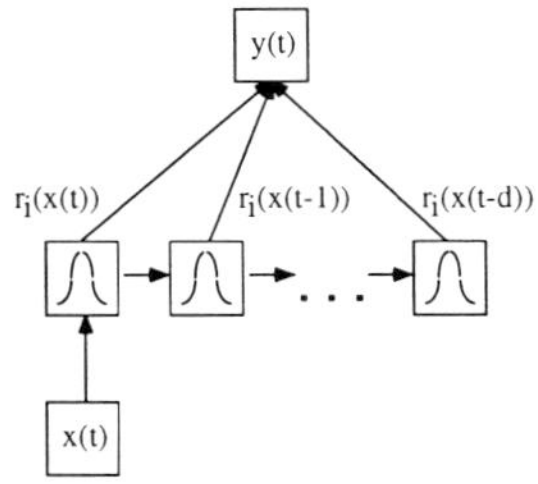

Figure 5. Structure of a Radial Basis Function Network with time delays in the hidden layer.

Figure 5 shows the structure of a RBFN network with d time-delays in the hidden (cluster) layer. The output neuron calculates its output y by adding current membership values and their *history* over the last *d* time steps, such that

$$y(t) = f(\vec{x}(t)) = s(\sum_{i=1}^{n}\sum_{j=0}^{d} w_{i,j} r_i(\vec{x}(t-j)))$$

In addition, it is also possible to delay the actual input vector $\vec{x}(t)$, i.e., to give each neuron access to $\vec{x}(t), \vec{x}(t-1), \ldots, \vec{x}(t-d)$ (Berthold, 1994). However, this increases the dimension of the input space from $dim(\vec{x})$ to $(d+1) \times dim(\vec{x})$ and, consequently, results in a much higher number of connections that are to be considered. Also, a higher number of examples is necessary to sufficiently cover the input space. Since the application investigated here is subject to real-time constraints, this option was not considered for the experiments described in Section 5.

3.4. *A Fuzzy Controller Architecture*

Fuzzy controllers are universal function approximators developed within the fuzzy logic community (Zadeh, 1992; Berenji, 1992). Even though they originate from a very different approach, they bear strong similarities to RBFNs and share the same learning algorithms. As a matter of fact, Fuzzy Controllers do not correspond to a single architecture but to a wide family that contains the topology of Figure 2b as a special case. Nevertheless, in our research, we focused on a specific topology, derived from the one proposed by Berenji (Berenji, 1992). Figure 6 shows the corresponding network representation.

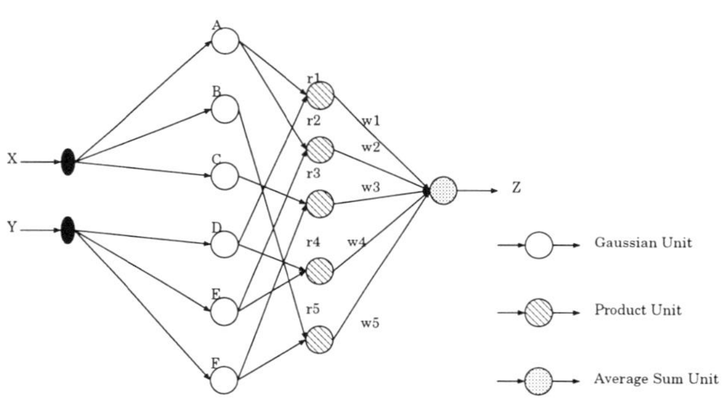

Figure 6. Network corresponding to the fuzzy controller used in current experiments. Gaussian units have a unidimensional Gaussian activation function. Product units compose the input values using arithmetic product or *softmin*. An average sum unit performs the weighted sum of the activation values received from the product units.

Differently than RBFNs, the fuzzy controller is organized as a three-layered network plus an input layer. The first hidden layer neurons are feature detectors with a unidimensional Gaussian activation field, which receives only one input signal. They compute the function

$$r_{ij} = e^{-\left(\frac{x_j - \mu_{ij}}{\sigma_{ij}}\right)^2} \tag{1}$$

where r_{ij} is the value of membership of the *j*-th component x_j of input $\vec{x}$ to a bell-shaped fuzzy set of center μ_{ij} and width σ_{ij}.

The neurons in the second hidden layer receive the membership values computed by two (or more) neurons of the previous layer, which are required to process different input signals. The output is obtained using the *softmin* composition rule

$$r_i = \frac{\sum_{j=1}^{dim(\vec{x})} r_{ij} e^{-kr_{ij}}}{\sum_{j=1}^{dim(\vec{x})} e^{-kr_{ij}}} \tag{2}$$

that tends to the regular min (the function usually employed for fuzzy AND) for $k \to \infty$, but is differentiable. Finally, the output neuron performs the linear combination

$$y = \frac{\sum_i w_i r_i}{\sum_i r_i} \tag{3}$$

of the weights w_i and the activation values r_i of the corresponding second layer neurons.

Comparing this structure to an RBFN, two differences can be noticed: first, the locally receptive field function is built as a combination of unidimensional functions using *softmin* (2) instead of product. This function is similar to a pyramid with smoothed corners and allows for a more precise symbolic interpretation. The second difference is the use of normalization (3) in the output unit. The consequence is an improved capability of generalization, so that good approximations can often be achieved using a smaller number of neurons than in a corresponding RBFN.

In the fuzzy set literature, the following symbolic interpretation is given to this architecture. The first hidden layer is interpreted as a mapping from continuous input features to linguistic variables (*Fuzzy Sets*) and is called *Fuzzifier*. The second hidden layer is seen as a set of implication rules in conjunctive form defining a logical mapping from the set of linguistic variables to a discretized set of output values. For instance, by referring to Figure 6, neuron r1 can be represented as a rule $A \cap D \to W_1$. Finally, the last layer is seen as a mapping from a discrete set of values to a continuous space and is called *Defuzzifier*.

4. Supervised Learning Algorithms for TDNN, RBFN and Fuzzy Controllers

Both in OFFNs and in LRFNs, layout construction and weight calculation are usually performed in separate steps. OFFNs' layout is usually defined in an empirical way by using rules of thumb to select both the number of neurons and the connnecting topology; algorithms such as KBANN (Towell et al., 1990) or Cascade-Correlation (Fahlmann & Lebiere, 1989) are designed for specific applications or architectures and do not help in the general case. On the contrary, several algorithms are available for automating LRFNs layout construction. The reason for this difference is to be sought in the *locality property* characterizing LRFNs; in fact, the hidden layer can be constructed by subdividing the input space into a mosaic of closed regions, allowing the approximation of the target function by means of a histogram. In RBFNs literature, this task is usually accomplished using clustering algorithms such as k-means or variants of it (Moody & Darken, 1989; Wilpon & Rabiner, 1985). Afterwards, the weights on the links to the output level can be learned either by performing a gradient descent of the quadratic error or using algorithms based on pseudo-inverse matrix transformation. Gradient descent can be quite slow but it can easily be adapted in order to be done incrementally on-line. On the contrary, the pseudo-inverse matrix transformation is fast but not incremental. Approaches to construct also the network layout in an incremental manner are, for instance, GAL (Alpaydin, 1991) and Fritzke's Growing Cell Structures (Fritzke, 1993).

The most common approach to design a Fuzzy Controller is to setup the network layout manually, relying on the domain knowledge of a human expert (Nuttin et al., 1994). Variants of the error gradient descent have been proposed in order to refine the fuzzy sets in a second step (Berenji, 1990).

In the framework of the present work, three alternative procedures have been defined in order to automate the layout construction of a LRFN. One of the procedures is based on a variant of the k-means according to the algorithm described in (Moody & Darken, 1989; Wilpon & Rabiner, 1985). A second procedure is based on CART (Breiman et al., 1984), an algorithm for generating Regression Trees. Finally, the third technique is based on symbolic learning methods in the line of (Sammut et al., 1992), and allows learning both from data and from background knowledge. In the present form, the procedure based on k-means is immediately applicable to the RBFN and TDRBFN architectures whereas the other two have been implemented in order to work on the architecture of the fuzzy controller. Nevertheless, they can be easily generalized to work indifferently on every kind of LRFN architecture. Afterwards, the networks can be trained on-line e.g. by error minimization using a gradient descent technique, which has the advantage that it is also applicable on-line.

4.1. Learning LRFNs' Layout by means of Statistical Clustering

The original proposal by Moody and Darken (Moody & Darken, 1988) for learning RBFNs' layouts was to use a modified k-means clustering algorithm to find a set of k clusters. However, k-means is an unsupervised algorithm, i.e., it does not take the output information into account. If the examples given for training the RBFN are in the form of $(\vec{x}, \vec{y})$ with $\vec{y}$ being the desired output vector for input $\vec{x}$, the network construction should use this information.

The procedure applied for the experiments described in Section 5 is based on work presented in (Musavi et al., 1992). Its purpose can be described as[3]:

Given: A set of examples $\{(\vec{x}_1, y_1), \ldots, (\vec{x}_n, y_n)\}$
Determine: A set of clusters $\{(\vec{\mu}_1, \sigma_1), \ldots, (\vec{\mu}_C, \sigma_C)\}, C \ll n$ that covers the given $\vec{x}_i$ and classifies them correctly with respect to the desired output y_i.

The algorithm used for initializing RBFNs networks is the following:

1. Divide the output interval $[y_{min}, y_{max}]$ into K equidistant and distinct intervals $I_1, \ldots, I_K$.

2. For all $i \in \{1, \ldots, n\}$: Define the index $k \in \{1, \ldots, K\}$ of the interval I_k covering y_i as the *class* of $\vec{x}_i$. This is denoted by $Class(\vec{x}_i) = k$.

3. Define a cluster $(\vec{\mu}, \sigma) := (\vec{x}_i, \beta\sigma_{min})$ for each $i \in \{1, \ldots, n\}$, with $\sigma_{min} := min_{i,j=1,i\neq j}^{n} ||\vec{x}_i - \vec{x}_j||$. This step results in $C = n$ initial clusters.
4. For all $i, j \in \{1, \ldots, C\}, i \neq j$, if $Class(\vec{\mu}_i) = Class(\vec{\mu}_j) = k$ try to merge the clusters:
   (A) Compute $\vec{\mu}_{new} : \mu_{new_l} = \frac{\mu_{i_l} + \mu_{j_l}}{2}$ for $l \in \{1, \ldots, dim(\vec{\mu}_i)\}$.
   (B) Compute $\sigma_{new} = \frac{1}{2}||\vec{\mu}_i - \vec{\mu}_j|| + max(\sigma_i, \sigma_j)$.
   (C) Compute $r = max_{l=1}^{n}\{||\vec{\mu}_{new} - \vec{x}_l|| : Class(\vec{x}_l) = k\}$.
   (D) Compute $d = min_{l=1}^{n}\{||\vec{\mu}_{new} - \vec{x}_l|| : Class(\vec{x}_l) \neq k\}$.
   (E) If $d > \alpha r$ remove cluster j, set $(\vec{\mu}_i, \sigma_i) := (\vec{\mu}_{new}, \sigma_{new})$. This means $C = C - 1$.
5. If at least two clusters have been merged (i.e., the number C of clusters has been reduced) go back to 4.

The parameters α and β, used in this procedure, have the following meaning: α tells how *conservative* the algorithm is about generating clusters that cover some example belonging to classes different than the one at issue. If $\alpha > 1$, no cluster covering wrong examples will be generated. If $\alpha < 1$, the algorithm is allowed to generate clusters covering wrong examples. Therefore, changing α is a way to influence both the accuracy that can be expected from the RBFN generated on the found clusters (a greater α means higher accuracy) and the size of the network (the smaller α, the more clusters will be merged and the less will be kept). Differently than the observations documented in (Musavi et al., 1992), for function approximation problems as they are considered here, selecting $\alpha < 1$ has proven to provide very satisfactory results, both in terms of network size and final approximation accuracy (see Section 5).

The β parameter determines the initial size of the clusters. A conservative selection for β is $\beta < 0.5$ or even $\beta = 0$, which guarantees that the initial clusters do not interfere. A setting of $\beta > 0.5$ or even $\beta > 1$ results in an initial cluster set that already provides ambiguous classifications. This is usually not desired.

4.2. *Learning LRFNs' Layouts with CART*

Regression Trees were introduced by (Breiman et al., 1984) for function approximation. A regression tree partitions the domain of a function $f(\vec{x})$ into rectangular regions (Fig. 7) where the value of $f(\vec{x})$ is similar, such that it can be approximated by a constant. Therefore, $f(\vec{x})$ is approximated by a histogram as accurately as the subdivision of the input domain permits.

The algorithm for generating a regression tree (CART) is simple and a detailed description can be found in (Breiman et al., 1984).

Given a learning set $\{(\vec{x}_1, y_1), \ldots, (\vec{x}_n, y_n)\}$, CART works as follows:

1. Start with a tree containing only one leaf, classifying all samples in the same category, and let the label of this leaf be the average value $\overline{f(\vec{x})}$;

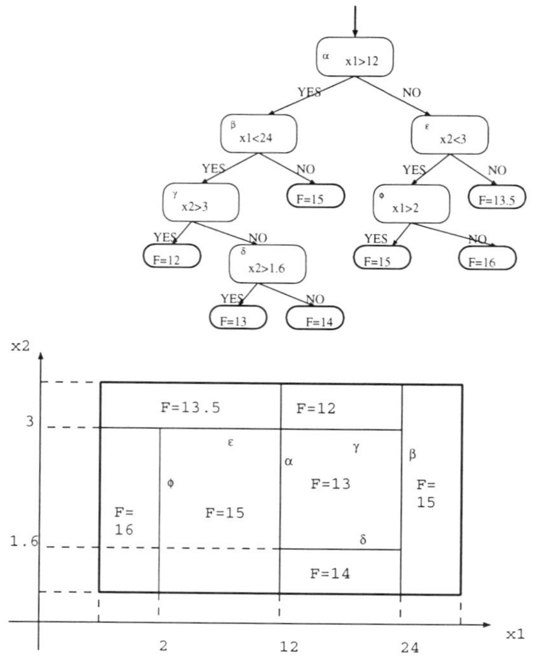

Figure 7. A Regression Tree

2. While the desired accuracy ϵ has not been achieved, repeat:

   (A) Choose a leaf F having an impurity $I(F)$ (the mean square error) greater than ϵ;

   (B) Find the feature x_j for which there exists a split value k_j so that the global impurity $I(F^{(1)}) + I(F^{(2)})$ for the two leaves $F^{(1)}$ and $F^{(2)}$, obtained by splitting F in k_j, is minimized;

   (C) Convert F into a node with the test $x_j \geq k_j$ and two leaves $F^{(1)}$ and $F^{(2)}$ as sons;

   (D) Label $F^{(1)}$ and $F^{(2)}$ with the average value of $f(\vec{x})$ in the corresponding regions of the domain.

To control overfitting, the regression tree can be pruned: the tree is explored bottom-up and every inner node (which corresponds to a test) is removed if this does not cause the error computed on a second learning set (the *pruning set*) to increase.

A regression tree can be employed directly to build a controller, but it has some major disadvantage: it approximates the function with a histogram (Fig. 10a), while smoothness of the output is desirable to avoid harmful mechanical solicitations to the robot; only its consequents (output values) can be fine-tuned, and not the input classification part, finally, the known learning algorithms for regression trees are not incremental, whereas we need an incremental tool for on-line refinement.

For all these reasons, we will convert a regression tree into the layout of a LRF network (both a Fuzzy Controller or a RBFN) that can the be further refined. This is done by the following algorithm:

1. For each leaf F_i in the tree, identify the rectangular region R_i it defines in the input space. This is done by collecting the tests on the path from the root to the leaf.

2. If the target network is a RBFN, every region R_i becomes a neuron r_i, centered on the center of R_i and spanning on it; in the case of a Fuzzy Controller, every edge σ_{ij} of R_i becomes a neuron in the first hidden layer, associated to a unidimensional Gaussian of amplitude σ_{ij}, connected to a neuron r_i in the second hidden layer and having the $softmin$ as activation function.

3. Every neuron r_i is connected to the output neuron by a link with weight w_i equal to the average value of $f(\vec{x})$ in the corresponding region R_i.

4.3. *Constructing LRFNs' Layouts using a Concept Learning Algorithm*

A regression problem can be mapped into a classification problem using a simple method: the real co-domain of the target function $f(\vec{x})$ is approximated by a finite set $\mathcal{O}$ of values. $f(\vec{x})$ is transformed into a histogram $f^*(\vec{x})$ by replacing each output value with the nearest value in $\mathcal{O}$. In this way, the regression problem becomes a classification problem, being $\mathcal{O}$ the set of classes and $f^*(\vec{x})$ the learning set, which can be solved using an algorithm such as C4.5. An example of this method is also described in (Sammut et al., 1992).

The classification knowledge in a decision tree can be processed similarly to the one encoded in regression trees, in order to obtain an LRFN network. The improvement, w.r.t. the work of Sammut and Michie, is that both fuzzy controllers and RBFNs can be further trained, obtaining a greater accuracy.

In the present case, we didn't use decision trees since they did not offer advantages over regression trees. Instead, we applied a more flexible learner called SMART+ (Bergadano et al., 1988; Botta & Giordana, 1993), that allows the induction process to be guided by the background knowledge of a domain expert. SMART+ is a complex system capable of learning in First Order Logics as well as in Propositional Calculus (as in the present case). A thorough description of the system is beyond the aim of this paper, thus we will only describe the subset of it that has been used here.

As well as FOIL (Quinlan, 1990) and FOCL (Pazzani & Kibler, 1992), SMART+ uses a general to specific learning strategy. Given a set of classes $\mathcal{O}$, it starts the search from a formula ϕ, which can be the predicate $True$ or a more complex formula (suggested by an expert or deduced from a domain theory) (Bergadano & Giordana, 1988).

One major feature, distinguishing SMART+ from other first order language learning systems, is its ability to deal with continuous attributes. This is done using predicate schemes such as

$$P(x_1, x_2, ..., x_n, K : [v_{min}, v_{max}, \delta]). \quad (4)$$

In (4) $x_1, x_2, ..., x_n$ are variables and K denotes a numerical threshold to be determined during the learning process. The interval $[v_{min}, v_{max}, \delta]$ defines the range for K and the required precision. Obviously, in a learning problem framed in propositional calculus, predicate schemes will not have any variable but only constants to tune. An example of a predicate used for learning a controller's layout can be $GreaterForce_x(K : [10, 25, 0.1])$ stating that the force along the X axis be greater than a constant K, ranging from 10 to 25 Newtons, with granularity 0.1.

In this case, the learning events are simply described as attribute vectors where $Force_x$ is one of them. The semantics of the predicate $GreaterForce_x$ is simply given by a function which compares the attribute $Force_x$ with the value assigned to K and returns true if the former is greater and false otherwise. SMART+ environment offers a function library designed in order to define the predicate semantics on the attributes of the learning events. In this way, background knowledge can be easily encapsulated in the predicate semantics.

A second method for supplying background knowledge is provided by necessity constraints which are statements of the type:

$$P(x_1, x_2, ..., x_n) \; needs \; \phi(y_1, y_2, ..., y_m) \tag{5}$$

where $\phi(y_1, y_2, ..., y_m)$ is an and/or formula belonging to the concept description language. The meaning of (5) is: $P(x_1, x_2, ..., x_n)$ can be true only if $\phi(y_1, y_2, ..., y_m)$ has been proven true. In general, SMART+ will use a predicate P for making a formula ψ more specific, only if its necessity constraint ϕ is a subformula of ψ. The induction space can be arbitrarily constrained by properly defining a set of necessity constraints.

The induction algorithm used by SMART+ for learning controller layouts learns one class o at a time:

Let $E(o_i)$ be the set of examples for the current target class o_i; let moreover, $\nu(\phi)$ be a measure of the quality of an inductive hypothesis ϕ. While $E(o_i)$ is not covered, or no new inductive hypotheses are found, do:

1. Start with an "or" tree T containing the formula "True", or a tree previously initialized with a set of formulas derived from a theory (or given by an expert).

2. Insert in the list $OPEN$ all the leaves currently existing in T.

3. Select from $OPEN$ the formula ϕ having the best ranking $\nu(\phi)$.

4. Determine the set P_A of predicates applicable to ϕ according to the predicate constraints.

5. For each predicate $p(K) \in P_A$, find the assignment k for the parameter K in p which maximizes $\nu(\phi \wedge p(k))$. If $\nu > 0$ put $\phi \wedge p(k)$ in the list $TODO$ sorted according to decreasing values of ν.

6. If $TODO$ is longer than a parameter $n \geq 1$, limiting the branching factor of T, truncate $TODO$ after the n-th item. Then, expand T with the formulas in $TODO$ and put them in $OPEN$.

7. If $OPEN$ contains some classification rule ψ(according to an assigned criterion), go to step 1; otherwise go to step 2.

The function ν used to rank an inductive hypothesis ϕ, has been chosen to be the function

$$if(v \geq v_0)\ then\ \nu = vw\ else\ \nu = 0 \tag{6}$$

with $0 \leq v \leq 1$ and $0 \leq w \leq 1$ being the proportion of positive examples covered by ϕ and the correctness of ϕ, respectively. The condition $v \leq v_0$ prevents the generation of too specific classification rules, in order to control the overfitting. Additionally, SMART+ offers a wide choice of evaluation criteria, including the well known $information\ gain$.

Inductive hypotheses are accepted as classification rules also if they are slightly inconsistent. In particular, a classification rule ϕ is allowed to cover some examples of the two classes corresponding to the two values in $\mathcal{O}$ adjacent to the target class o_i, provided that its correctness doesn't drop below an assigned threshold w_0. The tuning of w_0 and v_0 is left to the user and can require several trials to find a proper setting. A second constraint which can be posed on the acceptance of a classification rule is on the syntactic structure. In particular, it is possible to require such a rule to contain one or more predicates specified by the user. In this way, it is possible to achieve a complete characterization of the input region associated to the value represented by the target class.

The antecedents of the rules generated by SMART+ describe hyper-rectangles in the input space that can be processed as in the case of regression trees. Rule consequents are used for initializing the weights on the connections from the generated networks' hidden layer to the output neurons. Incompleteness and incorrectness of the rule set can easily be recovered by subsequently training the network's weights.

4.4. *Refining LRFNs On-line*

Let $E = \frac{1}{2}(Y - y)^2$ denote the square error exhibited by an approximation $y = f(\vec{x})$ of the function $Y = F(\vec{x})$ for a specific input $\vec{x}$. Let moreover $f(\vec{x})$ depend upon a set $\mathcal{P}$ of tunable parameters. The error gradient descent is performed by iteratively updating each parameter $P_i \in \mathcal{P}$ with the following rule (Rumelhart et al., 1985):

$$\Delta P_i = -\eta \frac{\partial E}{\partial P_i} = -\eta \frac{\partial E}{\partial y} \frac{\partial y}{\partial P_i} = \eta (Y - y) \frac{\partial y}{\partial P_i} \tag{7}$$

If the update of the parameter $\mathcal{P}$ takes place immediately, the learning will be characterized as *on-line*. If the effect is delayed until the whole learning set has been processed, the learning method is called *off-line* or *batch* learning. Batch learning guarantees more accurate results, and convergence proofs for backpropagation rely on batch learning as well. However, in robotics often on-line learning is required for obvious reasons.

In the following, we will derive the update rules for the different parameters of LRFNs, and we will show how they can be combined with other incremental learning rules.

4.4.1. *Refining RBFNs*

The typical RBFN training procedure is to consider the clusters $(\vec{\mu}, \sigma)$ as static. Only the weights between the cluster neurons and the output neuron are trained by means of the the rule (7) which for the case of $s = id$ simply reduces to: $\Delta w_i = \eta(Y - y)$. In the experimental section, networks trained using this simple method will be referred to as $RBFN_S$ or $TDRBFN_S$ (S stands for Simple).

Keeping the cluster centers constant during the whole life of the network is, however, not appropriate in all cases. Sometimes, not all data are available to perform the initial clustering. Also, the effort to analyze all given examples might be too high, thus only a subset considered representative is taken for setting up the network. A factor to be taken into special consideration w.r.t. real time applications is the size of the network, i.e. the time necessary for evaluation. The number of clusters should be small in such cases, making it much more difficult to yield the desired output accuracy by training the output weights only.

An obvious way to train the cluster centers during training is to apply a Kohonen-style nearest neighbour algorithm, also referred to as *adaptive incremental k-means* algorithm (Moody & Darken, 1988). Given an example $(\vec{x}, y)$, the center of the cluster closest to $\vec{x}$ is moved into the direction of $\vec{x}$ according to

$$\vec{x}_{closest} = \vec{x}_{closest} + \eta(\vec{x} - \vec{x}_{closest})$$

where η denotes a learning rate. During training, this adaptation can be interleaved with training the output weights, in order to avoid interference. Networks trained using this procedure will be referred to as $RBFN_{NN}$ and $TDRBFN_{NN}$.

Alternatively, if the cluster neurons' transfer function r is differentiable with respect to $\vec{\mu}$ and σ, it is possible to train both parameters by performing the error gradient descent, too (Weymaere & Martens, 1991; Wettschereck & Dietterich, 1991). Considering expression (7), for the case of a multidimensional Gaussian transfer function with individual σ_i for each dimension, i.e., for

$$r(\vec{x}) = e^{-(\sum_{i=1}^{dim(\vec{x})} (\frac{x_i - \mu_i}{\sigma_i})^2)},$$

the partial derivatives of y with respect to μ_i and σ_i are

$$\frac{\partial y}{\partial \mu_i} = 2\frac{\mu_i - x_i}{\sigma_i^2} r(\vec{x}) \quad \text{and} \quad \frac{\partial y}{\partial \sigma_i} = -2\frac{(\mu_i - x_i)^2}{\sigma_i^3} r(\vec{x})$$

In both cases, the error surface becomes extremely steep if the clusters are small, i.e., $\sigma_i \ll 1$. Since we are only interested in the sign of the gradient, we took care of this phenomenon by setting $\sigma_i = 1$ for the calculation of the partial derivatives. The networks trained using these methods are referred to as $RBFN_C$ or $TDRBFN_C$ and $RBFN_W$ or $TDRBFN_W$, respectively. If both methods are used in an interleaved way, this is referred to as $RBFN_{WC}$ or $TDRBFN_{WC}$.

4.4.2. *Refining a Fuzzy Controller via Back-Propagation*

The fuzzy controller of Figure 6 can be fine-tuned in a similar way as RBFNs; however, the $softmin$ composition function must be taken into account. When updating the weights in the last layer, equation (7) becomes

$$\Delta w_i = \eta(Y - y)\frac{r_i}{\sum_i r_i} \tag{8}$$

where $\frac{\partial y}{\partial w_i} = \frac{r_i}{\sum_i r_i}$ is the derivative of equation (3). For updating first layer weights (i.e., to adjust antecedent fuzzy sets), equation (7) must be rewritten as:

$$\Delta P_i = \eta(Y - y)\frac{\partial y}{\partial P_i} = \eta(Y - y)\sum_c \frac{\partial y}{\partial r_c}\frac{\partial r_c}{\partial \mu_i}\frac{\partial \mu_i}{\partial P_i} \tag{9}$$

where $\{P_i\} = \{C_i\} \cup \{\sigma_i\}$,

$$\frac{\partial y}{\partial r_s} = \frac{w_i\left(\sum_i r_i\right) - 1 \cdot \left(\sum_i w_i r_i\right)}{\left(\sum_i r_i\right)^2} \tag{10}$$

is the derivative of (3),

$$\frac{\partial r_c}{\partial \mu_i} = \frac{\frac{\partial \mathcal{N}}{\partial \mu_i}\mathcal{D} - \frac{\partial \mathcal{D}}{\partial \mu_i}\mathcal{N}}{\mathcal{D}^2} \tag{11}$$

$$\left(\mathcal{D} = \sum_l e^{-k\mu_l} \quad ; \quad \mathcal{N} = \sum_j \mu_j e^{-k\mu_j} \quad ; \quad r_c = \mathcal{N}/\mathcal{D} \quad \text{for brevity}\right)$$

is the derivative of the *softmin* operator (2) and

$$\frac{\partial \mu_i}{\partial \sigma_i} = \frac{2\mu_i(x - C_i)^2}{\sigma_i^3} \quad ; \quad \frac{\partial \mu_i}{\partial C_i} = \frac{2\mu_i(x - C_i)}{\sigma_i^2} \tag{12}$$

are the derivatives of the Gaussian membership functions (1).

5. Experimental Evaluation

In the following we will compare the function approximators and the learning procedures proposed in Sections 2 and 4, on two test cases. The first is a robotic application, in which the task is to approximate the output of a controller that is already operational on a robot manipulator. The goal of the experiments was to check the accuracy of the methods in capturing control functions. The results proved that all the approximators are potentially able to capture the control function from examples of correct behavior only.

The second test case is a chaotic function, the Mackey-Glass temporal series, used by many authors to check the predictivity of a function approximator.

What emerges from the experiments is that LRFNs are in general more accurate, especially when they are synthesized using the symbolic approaches based on CART and on SMART+.

5.1. *Experiments on the Robot Traces*

In this case, the experimental test-bed was given by a six-degrees-of-freedom KUKA-IR 361 manipulator equipped with a force-torque sensor. As the robot is kinematically controlled, the controller generates a vector $\vec{V}$ of translational and rotational velocities of the wrist.

The selected task, *peg-into-hole*, consists in learning to insert a peg into a hole and to recover from error situations, in which, for instance, the peg is stuck midway because of a wrong inclination. This application is particularly interesting because the optimal control is known to be non-linear (Asada, 1990).

In our case, both the peg and the hole had a circular section. The diameter of the chamfered peg was $30[mm]$, the clearance between the peg and the hole was $0.15[mm]$. The hole was located on a plane surface. In the experiments, a PID-controller was already available for the robot (De Schutter & Van Brussel, 1988), and the goal was to learn to approximate the behavior of the PID-controller using a set of examples of control behavior.

These examples were obtained by recording the traces of the pairs of input sensors readings and corresponding speed values, during seven insertions with different initial conditions such as different misalignments between the peg and the hole. Three of these sequences were chosen as learning examples, the others were used as the test set. Figure 8 shows the input signals for the seven sequences. Figure 9 shows the three output signals for the linear velocity, generated by the PID-controller and used to train our function approximators.

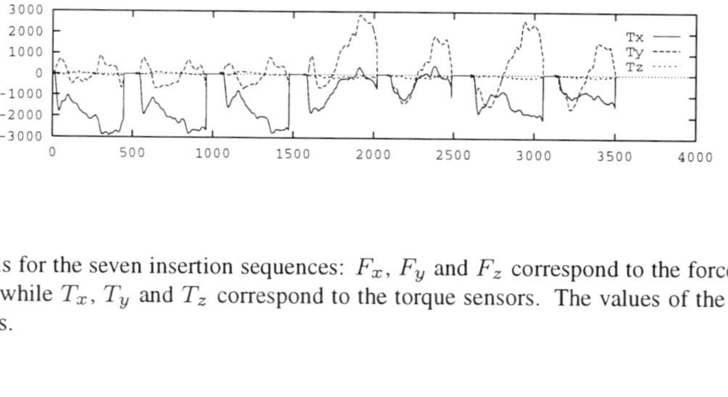

Figure 8. Input signals for the seven insertion sequences: F_x, F_y and F_z correspond to the force sensors along the three coordinates while T_x, T_y and T_z correspond to the torque sensors. The values of the functions have been sampled at 10 ms.

5.1.1. *Comparative Results on Robot Traces*

All learning methods have been applied to the case study, using the same learning and test sets described above. In order to make comparable the results, the refinement performing the error gradient descent has been done on-line and stopped after 1,500,000 learning steps.

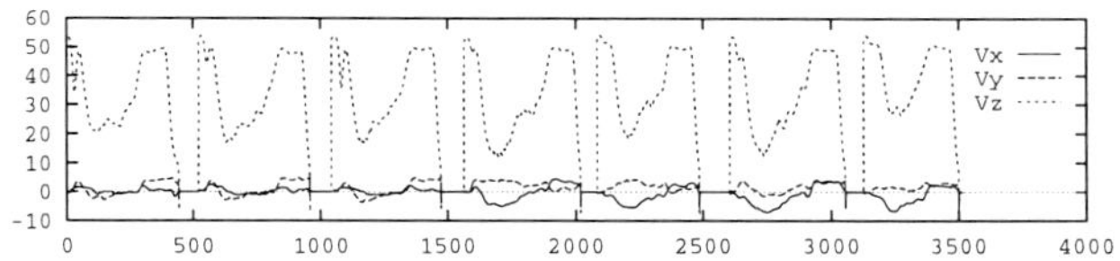

Figure 9. Velocity signals computed by the PID-Controller from the signals reported in the previous figure.

The method based on CART turned out very easy to apply because it didn't require any special setting on control parameters, while the one based on SMART+ required a little more work in order to decide the subdivision of the co-domain of the target functions into classes. After several trials it turned out that a subdivision into 12 equal intervals was a good choice for all three function V_x, V_y and V_z. Moreover, a set of predicate constraints has been defined in order to force SMART+ to generate rules necessarily accounting for the three features V_x, V_y and V_z. The complexity of the trees generated by CART and SMART+ was comparable and ranged from 20 to 80 leaves corresponding to a number of neurons in the hidden layer ranging from 60 to 160 (each neuron corresponds to an unidimensional Gaussian).

RBFNs and TDRBFNs were constructed using the algorithm described in section 4.1 and have been trained using the alternative algorithms we described in section 4. For the TDRBFN architecture only 1 delay has been used everywhere. The complexity of the networks ranged from 30 to 60 neurons in the hidden layer (each neuron corresponds to a 6 dimensional Gaussian).

For the TDNN experiments, the structure of the network and the number of delays at each layer was designed manually in a trial-and-error manner. Afterwards, the networks were trained by backpropagation (Rumelhart et al., 1985). The topologies used had a complexity ranging from 10 to 20 neurons. The best number of delay units found during the experimentation was 3 in the input layer and 0 in the other layers.

Finally, the experiment has been done using the basic multilayer perceptron for comparison purposes. Also in this case, the choice of the layout required several experiments.

The results are reported in Table 1. It is easy to see the methods based on CART and SMART+ obtained very good results (CART is the best in absolute) as well as the TDNN and the TDRBFN. On the contrary multilayer perceptron and normal RBFN were not so good even if in general they can achieve acceptable results.

Concerning the training algorithm for RBFN we notice the relevance of training the hidden layer neurons. In particular, the methods based on gradient descent performed better than the one based on incremental k-means.

In order to better understand the behavior of the two methods based on CART and SMART+, some details of the learning process are illustrated in Figure 10 and in Figure 11.

In particular, Figure 10a shows how CART alone approximates the target function with a histogram; Figure 10b shows the output for the same sequence generated by the fuzzy

Table 1. Statistical comparative results; all numbers are percentages (%)

System	V_x $Err_{avg} \pm Std.dev.$	V_y $Err_{avg} \pm Std.dev.$	V_z $Err_{avg} \pm Std.dev.$
CART (alone)	0.76 ± 1.05	1.02 ± 1.33	1.38 ± 5.01
FC_{CART}	1.16 ± 2.51	1.77 ± 3.75	11.84 ± 20.92
FC_{CART} + R	0.04 ± 0.17	0.05 ± 0.31	0.45 ± 2.75
FC_{SMART+}	1.49 ± 2.02	1.28 ± 1.93	2.12 ± 4.94
FC_{SMART+} + R	0.73 ± 0.42	0.56 ± 0.35	1.04 ± 4.18
$RBFN_S$	3.0 ± 0.61	2.4 ± 0.37	3.4 ± 3.81
$TDRBFN_S$ (1 TD)	6.2 ± 0.85	11.5 ± 1.11	2.6 ± 2.37
$RBFN_{NN}$	2.9 ± 0.67	7.6 ± 0.78	4.5 ± 4.62
$TDRBFN_{NN}$ (1 TD)	2.4 ± 0.45	1.8 ± 0.18	1.6 ± 3.15
$RBFN_C$	2.2 ± 0.45	3.6 ± 0.38	3.3 ± 3.80
$TDRBFN_C$ (1 TD)	1.7 ± 0.32	1.7 ± 0.18	1.0 ± 3.00
$RBFN_W$	2.1 ± 0.40	2.9 ± 0.29	3.3 ± 3.83
$TDRBFN_W$ (1 TD)	1.9 ± 0.29	0.8 ± 0.09	1.7 ± 3.10
$RBFN_{CW}$	2.4 ± 0.50	2.3 ± 0.33	2.5 ± 3.67
$TDRBFN_{CW}$ (1 TD)	1.5 ± 0.25	1.4 ± 0.15	1.9 ± 3.15
TDNN	3.9 ± 0.59	8.1 ± 0.90	2.7 ± 2.30
MLP	7.9 ± 1.13	10.9 ± 1.10	4.5 ± 3.84

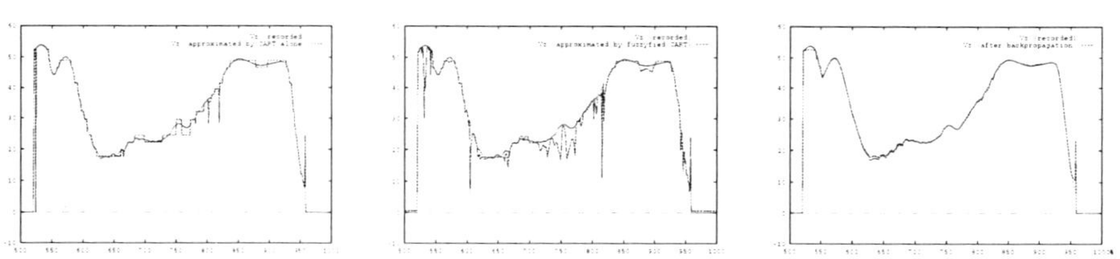

(a) (b) (c)

Figure 10. Approximation of the control function V_z obtained using: (a) CART only; (b) after transforming the tree of CART into a fuzzy controller; (c) after training the fuzzy controller performing the error gradient descent.

controller produced by converting the CART tree into a set of fuzzy rules. Figure 10c shows the signal after further training with backpropagation. The signal shown is obtained from the second training example and is the velocity V_z.

In a similar way, the procedure based on SMART+ is illustrated in Figure 11. In particular, 11a reports the histogram approximating the signal used as a learning set. Figure 11b shows the approximation obtained by the fuzzy controller before the training and 11c shows the result after the refinement step.

5.1.2. Results on the Robot Simulator

In order to test the learnt controllers "on the field", we used a professional robot simulation package (De Schutter et al., 1993).

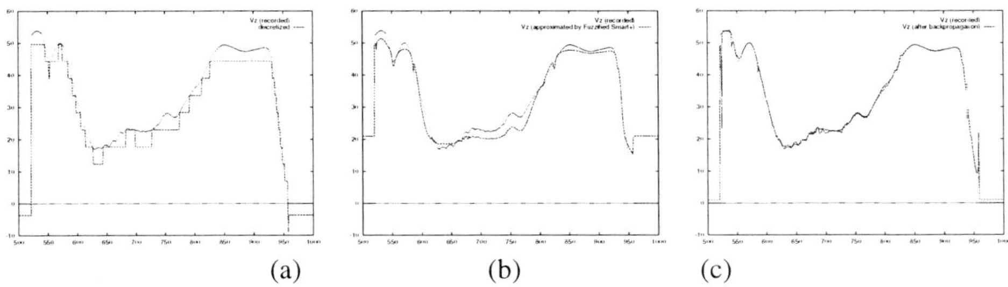

Figure 11. Approximation of the control function V_z obtained using SMART+: (a) histogram representing one of the signals in the learning set; (b) approximation generated by the fuzzy controller generated by SMART+; (c) after training by performing the error gradient descent.

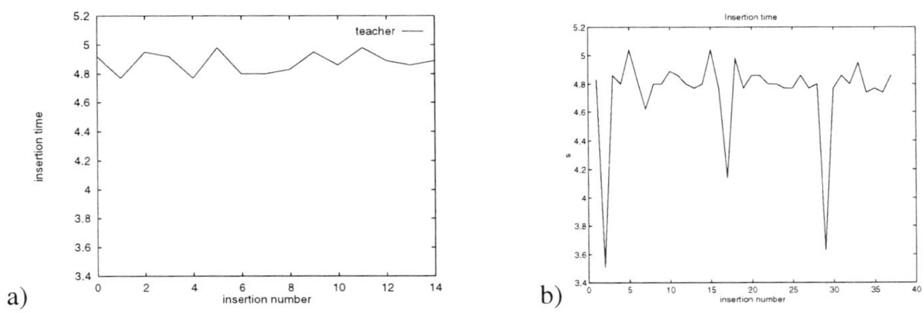

Figure 12. The behavior of the robot with the teacher (a) and with the learnt controller (b)

Figure 12 shows the behavior of the simulated robot arm: every trace depicts a series of insertions; each insertion is evaluated with the total insertion time (the lower, the better). As can be seen, the learnt controller exhibits a behavior comparable to the one of the teacher.

However, a test on the real robot should be more convincing. This has not been done for the simple reason that a controller generated using heuristic methods was not considered safe enough for a robot delicate and expensive as the KUKA IR-361 is. This fact points out another critical point which must be faced before using neural controllers in industrial applications, i.e., the development of validation techniques that are accepted by the mechanical engineers (ERA, 1995).

5.2. *Test on Mackey-Glass Chaotic Time Series*

As the learning procedures based on CART and SMART+ represent an important novelty which performed very well on robot traces, they have been tested also on a classical case study widely used in the literature in order to have a more extensive validation and a comparison with other methods. The task is to predict the value 84 time steps ahead in the

Mackey-Glass chaotic series, which is described by the following differential equation:

$$\dot{x} = \frac{0.2x(t-\tau)}{1+x^{10}(t-\tau)} - 0.1x(t)$$

with $x(t<0)=0$, $x(0)=1.2$ and $\tau=17$; the task was to predict $x(t+84)$ given $x(t-24)$, $x(t-18)$, $x(t-12)$ and $x(t-6)$.

The input features used by the approximators are the four values of the function at time $t, t-6, t-12$, and $t-18$. In order to correctly predict the function value, it is necessary to capture the generative model of the phenomenon. This task was also investigated by (Crowder, 1990; Jang, 1993; Jones et al., 1990; Lapedes & Farber, 1987; Moody, 1989; Moody & Darken, 1989; Sanger, 1991), so results can be compared directly.

The experiment has been organized as follows. First, a sequence of 1500 time steps has been generated. Then the learning set has been obtained by taking the first 1000 whereas the remaining 500 have been used as a test set. The results are reported in Table 2. The Non-Dimensional Error Index (NDEI) is defined as the Root Mean Square Error divided by the Standard Deviation of the target series.

Comparing the best results obtained in the literature in analogous experiments (Crowder, 1990; Jang, 1993), it appears that both the predictor generated by CART and the one generated by SMART+ show excellent performances after training. In fact, the NDEI value of 0.036 reported by ANFIS, followed by the 0.05 of the multilayer perceptron are the only performances really comparable to those presented here.

However, the big advantage of our new methods is that the networks have been synthesised automatically from examples, whereas for both ANFIS and the multilayer perceptron this has been done manually.

Regarding the learning procedure based on modified k-means, we refer to the work by (Moody & Darken, 1989). He reports a NDEI of 0.055 obtained with a training set of 10,000 examples and 1,000 hidden neurons. The method based on SMART+ obtained a better result using a fuzzy controller of 91 rules corresponding to a global number of 305 neurons in the first hidden layer. Moreover, the learning set was of only 1,000 examples.

The controller generated by CART was even smaller: 42 leaves in the tree corresponding to a global number of 161 neurons in the first hidden layer.

Table 2. Comparative results on the Mackey-Glass chaotic time series prediction

Method	Training cases	Non-Dimensional Error Index
CART (Alone)	800+200	0.36
FC_{CART} + R	1000	0.037
FC_{SMART+} + R	1000	0.032

6. Discussion

From the results presented in the previous section the excellent approximation capability of LRFNs clearly emerges. In fact, they obtained favourable results w.r.t. the multilayer perceptron and TDNNs, both on robot traces and on the Mackey-Glass series.

It is worth noticing that, considering the aim of robotic applications, the slight differences of performance between OFFNs and LRFNs (though certainly meaningful from a statistical viewpoint) cannot be decisive when looking for a technique to apply. Differences in the third decimal digit are not a sufficient reason for adopting a TDNN rather than an RBFN or a fuzzy controller; it is more fair to assume OFFNs and LRFNs as equivalent with respect to the accuracy.

Other considerations must, therefore, be taken into account in order to allow to select one architecture over another. LRFNs have specific properties that make them particularly suitable for robotic applications. The first is *support of incremental learning*, which is due to the locality property and has been discussed in Section 3. The possibility of incrementally building the mapping from the input space to the output space is equivalent to split the learning problem into subproblems; thus, different parts of the control function can be learnt during different learning sessions. For example the robot can first be trained to work slowly. Afterwards, specific experiments can be performed in order to increase its speed. The locality property also helps to avoid the *unlearning problem*, that arises in OFFNs when trained for a long time on a subset of the input space only: the network overspecializes and forgets knowledge about regions of the input space it doesn't see for a while. Although we didn't directly address this aspect of incrementality in our work, several examples can be found in the literature (see for instance (del R. Millán & Torras, 1992; del R. Millán, 1994)).

A second important feature is the possibility of giving a *symbolic interpretation* to LRFNs. We exploit this property when we apply symbolic and statistical learning algorithms for producing the network layout. In our opinion, this is a key point of the presented work, that establishes a *bridge* between approaches that (with a few exceptions) have traditionally been considered far apart, and thus scarcely interacted. Beyond the theoretical interest, we especially see practical benefits in the capability of merging the knowledge of an expert into a network.

In the literature, at least two other methods have already been proposed in order to integrate the symbolic and connectionist paradigms. The first is KBANN by Shavlik and Towell (Towell et al., 1990), who proposed to use a propositional theory in order to initialize a multilayer perceptron. The considered task was classification in a domain of Boolean features. The experiments showed that if a theory, which is a good approximation of the target Boolean function, is available, learning itself speeds up dramatically and the classification rate increases as well.

The main difference between the approach proposed in this paper and that of Shavlik's depends on the task they tackle. Learning a Boolean classifier is a simpler problem than learning a continuous function in a continuous domain. In this second case, the problem of learning the mapping from input signals to symbols, and then from symbols to an output signal, is also included in the process. Such mappings, which are naturally supported by the RBFNs are left out from the learning algorithm used by KBANN.

KBANN has been improved by adding algorithms for turning a multilayer perceptron back into a propositional theory (see (Towell & Shavlik, 1993) for a description of the method), a process we will call "inverse mapping of the network".

Inverse mappings can be found, in general, also in the case of LRFNs, for two reasons: (1) because of the locality property, that allows a discretization of inputs and outputs; (2) because they maintain their topological structure during learning. In particular, in the case of RBFNs and FCs, network inverse mappings can be obtained simply by reversing the process of encoding the theory into a neural net. Last, for the sake of completeness, it is necessary to pinpoint a difference between the kind of theories we can map onto the layout of an LRFN and those used by KBANN: the former are flat, in the sense that consequents of rules are not used in antecedents of other rules, whereas the latter can be structured into many layers.

A last point, needing a more detailed discussion, is the necessity of continuing the refinement of a controller beyond the teacher performances. Especially in the case of robot control, it is quite unlikely that a human operator will generate perfect examples of behavior (Kaiser et al., 1995c; Kaiser et al., 1995a). Therefore, some other solution has to be sought in order to optimize the induced controller, and Reinforcement Learning seems to be a promising approach. Actually, in RL literature we find many techniques for learning in continuous domains (Peng & Williams, 1992; Berenji & Khedkar, 1992; Williams, 1992; Gullapalli, 1990), in which agents implemented as neural networks are trained by means of variants of backpropagation. In all these cases, the ability to learn from a reinforcement signal is achieved by finding a quantity (depending on it) to backpropagate, i.e. playing the part played by the error in supervised learning rules. Most of the techniques proposed are, for instance, related to the TD(λ) method, in which differences of prediction in subsequent timesteps are exploited (see (Sutton & Barto, 1987)).

One system that integrates neural and reinforcement learning is Berenji's GARIC (Berenji & Khedkar, 1992). This system uses a fuzzy controller, implemented as a neural network, which is very similar to those described in Section 3. However, it is extremely easy to think to GARIC's variants in which the other kinds of LRFN approximators are used instead of fuzzy controllers. Therefore, the supervised learning methodologies presented here can be thought of as building blocks for more complex architectures.

7. Conclusions

Throughout this paper, we presented results achieved in the framework of B-LEARN II, a project that investigates the suitability of ML techniques for developing "advanced" industrial robots. Nonetheless, the fundamental claims that have been presented (supported by an extensive experimentation), are interesting also for ML researchers.

The first claim is that there is a class of problems related to non-linear control that can potentially be solved by employing existing learning algorithms. More specifically, the application of ML techniques to such problems can lead to real industrial applications, with an increase in robot performance and a decrease in costs of controller design and development. In particular, the experiments presented in this paper confirm that many techniques, originating from different areas, are available for facing the regression problem tied to non-linear control.

The second important claim concerns the integration of the symbolic and connectionist paradigms. We have shown that, on the way opened by Shavlik and Towell (Towell et al.,

1990; Towell & Shavlik, 1993), it is possible to use symbolic knowledge in order to build a quite accurate approximator that can further be improved using, e.g., backpropagation. This result opens a wide perspective for the integration of symbolic and non-symbolic methods in regression tasks.

At the same time we also established a link between ML and a parallel field, Fuzzy Logic, in which function approximators have also been developed combining symbolic and non-symbolic paradigms.

We made the attempt of adopting a unifying view for comparing architectures developed by different communities with, sometimes, competing approaches. We believe we have shown that, beyond the cultural background, there is a lot in common among them, which *must* be taken into account.

Notes

1. This *is* to be considered an advanced technological solution, since today's industrial robots make a very little use of external sensors.
2. **P**roportional **I**ntegral **D**ifferential controller, i.e., a controller which produces an output that is proportional to its input, the input derivative, and the input accumulated over time. The input might be the error, e.g., the difference between the desired force and the actually measured force.
3. For clarity, only a one dimensional output vector is considered. However, the method works on examples featuring arbitrary dimensions. With $dim(\vec{y}) = n$, the intervals I_j are becoming hypercubes of dimension n, and the number of classes to be considered becomes $n \times K$.

References

Alpaydin, E. (1991). GAL: Networks that grow when they learn and shrink when they forget. Technical Report TR-91-032, International Computer Science Institute, Berkeley, USA.

Asada, H. (1990). Teaching and learning of compliance using neural nets: Representation and generation of nonlinear compliance. In *Proceedings of the 1990 IEEE International Conference on Robotics and Automation*, pages 1237 – 1244.

Barto, A., Sutton, R., & Watkins, C. (1990). Sequential decision problems and neural networks. In *Advances in neural information processing system*, volume 2. Morgan Kauffman, San mateo, Ca.

Barto, A. G., Sutton, R. S., & Anderson, C. W. (1983). Neuronlike elements that can solve difficult learning control problems. *IEEE Transactions on Systems, Man and Cybernetics*, pages 835–846.

Berenji, H. (1990). Machine learning in fuzzy control. In *International Conference on Fuzzy Logic & Neural Networks*, pages 231–234, Iizuka, Fukuoka, Japan.

Berenji, H. (1992). Fuzzy logic controllers. In Yager, R. and Zadeh, L., editors, *An Introduction to Fuzzy Logic Applications in Intelligent Systems*. Kluwer Academic Publishers.

Berenji, H. & Khedkar, P. (1992). Learning and tuning fuzzy controllers through reinforcements. *IEEE Transactions on neural networks*, 3(5):724–740.

Bergadano, F. & Giordana, A. (1988). A knowledge intensive approach to concept induction. In *Proceedings of the 5th International Conference on Machine Learning*, pages 305–317, Ann Arbor, MI. Morgan Kauffman.

Bergadano, F., Giordana, A., & Saitta, L. (1988). Learning concepts in noisy environment. *IEEE Transaction on Pattern Analysis and Machine Intelligence*, pages 555–578.

Berthold, M. (1994). A time delay radial basis function network for phoneme recognition. In *IEEE International Conference on Neural Networks*, Orlando, Florida.

Bonissone, P. & Chiang, K. (1993). Fuzzy logic controllers: from development to deployment. In *IEEE International Conference on Neural Networks*, volume 2, San Francisco, CA.

Botta, M. & Giordana, A. (1993). SMART+: A multi-strategy learning tool. In *IJCAI-93, Proceedings of the Thirteenth International Joint Conference on Artificial Intelligence*, volume 2, Chambéry, France.

Breiman, L., Friedman, J., Ohlsen, R., & Stone, C. (1984). *Classification And Regression Trees*. Wadsworth & Brooks, Pacific Grove, CA.

Cramer, H. (1974). *Mathematical Methods of Statistics*. Princeton University Press.

Crowder, R. (1990). Predicting the mackey-glass time series with cascade-correlation learning. In D. Touretzky, G. H. and T.Sejnovsky, editors, *Proceedings of the 1990 Connectionist Models Summer School*, pages 117–123. Carnegie Mellon University.

De Schutter, J. & Van Brussel, H. (1988). Compliant robot motion II, a control approach based on external control loops. *The International Journal of Robotics Research*, 7(4).

De Schutter, J., Witvrouw, W., Van De Poel, P., & Bruyninckx, H. (1993). Rosi: a task specification and simulation tool for force sensor based robot control. In *24th International Symposium on Industrial Robots*.

del R. Millán, J. (1994). Learning efficient reactive behavioral sequences from basic reflexes in a goal-directed autonomous robot. In *Proceedings of the third International Conference on Simulation of Adaptive Behavior*.

del R. Millán, J. & Torras, C. (1992). A reinforcement connectionist approach to robot path finding in non-maze-like environments. *Machine Learning*, 8:363–395.

ERA (1995). *Neural Networks: Producing Dependable Systems*, Solihull, West Midlands, UK. ERA Technology.

Fahlmann, S. E. & Lebiere, C. (1989). The cascade-correlation learning architecture. In *Advances in Neural Information Processing Systems 2 (NIPS-2)*, Denver, Colorado.

Fritzke, B. (1993). Growing cell structure: A self-organizing network for unsupervised and supervised learning. Technical Report TR-93-026, International Computer Science Institute.

Gullapalli, V. (1990). A stochastic reinforcement learning algorithm for learning real valued functions. *Neural Networks*, 3:671–692.

Hornik, K., Stinchcombe, M., & White, H. (1989). Multilayer feed-forward networks are universal approximators. *Neural Networks*, 2:359–366.

Jang, J. (1993). ANFIS: Adaptive-Network-Based Fuzzy Inference System. *IEEE Transactions on Systems, Man and Cybernetics*, SMC-23(3):665–687.

Jones, R., Lee, Y., Barnes, C., Flake, G., Lee, K., & Lewis, P. (1990). Function approximation and time series prediction with neural networks. In *Proceedings of IEEE International Joint Conference on Neural Networks*, pages I–649–665.

Kaiser, M., Camarinha-Matos, L., Giordana, A., Klingspor, V., del R. Millàn, J., Nuttin, M., & Suarez, R. (1994). Robot learning - three case studies in robotics and machine learning. In *Proceedings of the IVAR '94*, Leuven, Belgium.

Kaiser, M., Friedrich, H., & Dillmann, R. (1995a). Obtaining good performance from a bad teacher. In *International Conference on Machine Learning, Workshop on Programming by Demonstration*, Tahoe City, California.

Kaiser, M., Klingspor, V., del R. Millàn, J., Accame, M., Wallner, F., & Dillmann, R. (1995b). Using machine learning techniques in real-world mobile robots. *IEEE Expert*.

Kaiser, M. & Kreuziger, J. (1994). Integration of symbolic and connectionist processing to ease robot programming and control. In *ECAI'94 Workshop on Combining Symbolic and Connectionist Processing*, pages 20 – 29.

Kaiser, M., Retey, A., & Dillmann, R. (1995c). Robot skill acquisition via human demonstration. In *Proceedings of the International Conference on Advanced Robotics (ICAR '95)*.

Lapedes, A. & Farber, R. (1987). Nonlinear signal processing using neural networks: Prediction and system modeling. Technical Report LA-UR-87-2662, Los Alamos National Laboratory.

Mason, M. (1981). Compliance and force control for computer controlled manipulators. *IEEE Transactions on Systems, Man and Cybernetics*, 11.

Miller, W. T., Sutton, R. S., & Werbos, P. J. (1990). *Neural networks for control*. The MIT Press.

Moody, J. (1989). Fast learning in multi-resolution hierarchies. In Touretzky, D., editor, *Advances in Neural Information Processing*. Morgan Kaufmann.

Moody, J. & Darken, C. (1988). Learning with localized receptive fields. In Sejnowski, T., Touretzky, D., and Hinton, G., editors, *Connectionist Models Summer School*, Carnegie Mellon University.

Moody, J. & Darken, C. (1989). Fast learning in networks of locally tuned units. *Neural Computations*, 1(2):281–294.

Musavi, M., Ahmed, W., Chan, K., Faris, K., & Hummels, D. (1992). On the training of radial basis function classifiers. *Neural Networks*, 5:595–603.

Nuttin, M., Van Brussel, H., Baroglio, C., & Piola, R. (1994). Fuzzy controller synthesis in robotic assembly: Procedure and experiments. In *FUZZ-IEEE-94: Third IEEE International Conference on Fuzzy Systems, World Congress on Computational Intelligence*.

Nuttin, M., Van Brussel, H., Peirs, J., Soembagijo, A. S., & Sonck, S. (1995). Learning the peg-into-hole assembly operation with a connectionist reinforcement technique. In *Second International CIRP Workshop on Learning in Intelligent Manufacturing Systems*, pages 335–357, Budapest, Hungary.

Park, J. & Sandberg, W. (1993). Universal approximation using radial-basis functions. *Neural Computation*, 5.

Pazzani, M. & Kibler, D. (1992). The utility of knowledge in inductive learning. *Machine Learning*, 9:57–94.

Peng, J. & Williams, R. (1992). Efficient learning and planning within the Dyna framework. In *Proceedings of the Second International Conference on Simulation of Adaptive Behavior*, Honolulu, HI.

Poggio, T. & Girosi, F. (1990). Networks for approximation and learning. *Proceedings of the IEEE*, 78(9):1481–1497.

Quinlan, J. (1993). Combining instance-based and model-based learning. In *Proceedings of the* 10th *machine learning conference*, pages 236–243, Amherst, MA.

Quinlan, R. (1990). Learning logical definitions from relations. *Machine Learning*, 5:239–266.

Rumelhart, D., Hinton, G., & Williams, R. (1985). Learning internal representations by error propagation. Technical Report 8506, Institute for Cognitive Science, La Jolla: University of California, San Diego.

Rumelhart, D. E. & McClelland, J. L. (1986). *Parallel Distributed Processing : Explorations in the Microstructure of Coginition, Parts I & II*. MIT Press, Cambridge, Massachusetts.

Sammut, C., Hurst, S., Kedzier, D., & Michie, D. (1992). Learning to fly. In Sleeman, D. and Edwards, P., editors, *Machine Learning - Proceedings of the Ninth International Workshop (ML92)*, pages 385–393. Morgan Kaufmann.

Sanger, T. (1991). A tree-structured adaptive network for function approximate in high-dimensional spaces. *IEEE Transactions on Neural Networks*, 2(2):285–293.

Specht, D. (1988). Probabilistic neural networks for classification mapping, or associative memory. In *IEEE International Conference on Neural Networks*, volume 1, pages 525–532.

Specht, D. (1990). Probabilistic neural networks. *Neural Networks*, 3:109–118.

Sutton, R. & Barto, A. (1987). A temporal-difference method of classical conditioning. In *proceedings of the Ninth Annual Conference of the Cognitive Science Society*, pages 355–378, Seattle, WA. Lawrence Erlbaum.

Towell, G. & Shavlik, J. (1993). Extracting refined rules from knowledge-based neural networks. *Machine Learning*, 13(1):71–101.

Towell, G., Shavlik, J., & Noordwier, M. (1990). Refinement of approximate domain theories by knowledge-based neural networks. In *Proceedings of the* 8th *National Conference on Artificial Intelligence AAAI'90*, pages 861–866.

Waibel, A., Hanazawa, T., Hinton, G., Shikano, K., & Lang, K. (1989). Phoneme recognition using time-delay neural networks. *IEEE Transactions on acoustics, speech and signal processing*, pages 328–339.

Wettschereck, D. & Dietterich, T. (1991). Improving the performance of radial basis function networks by learning center locations. In *Advances in Neural Information Processing Systems 4 (NIPS-4)*.

Weymaere, N. & Martens, J. (1991). A fast and robust learning algorithm for feedforward neural networks. *Neural Networks*, 4.

Williams, R. J. (1992). Simple statistical gradient-following algorithms for connectionist reinforcement learning. *Machine Learning*, pages 229 –256.

Wilpon, J. & Rabiner, L. (1985). A modified k-means clustering algorithm for use in isolated work recognition. *IEEE transactions on acoustics, speech and signal processing*, ASSP-33:587–594.

Yih, J. & Shieh, J. (1992). On the development of a fuzzy model-based controller for robotic manipulators. In *Proceedings of the IEEE/RSJ Conference on Intelligent Robots and Systems*, Raleigh, NC.

Zadeh, L. (1992). Knowledge representation in fuzzy logic. In Yager, R. and Zadeh, L., editors, *An Introduction to Fuzzy Logic Applications in Intelligent Systems*. Kluwer Academic Publishers.

Machine Learning, 23, 251–278 (1996)

Active Learning for Vision-Based Robot Grasping

MARCOS SALGANICOFF salganic@asel.udel.edu
Applied Science and Engineering Labs, University of Delaware/Alfred I. duPont Institute, P.O.B. 269, Wilmington, DE 19899

LYLE H. UNGAR ungar@central.cis.upenn.edu
RUZENA BAJCSY bajcsy@central.cis.upenn.edu
Department of Computer and Information Science, University of Pennsylvania, Philadelphia, PA, 19104

Editors: Judy A. Franklin, Tom M. Mitchell, and Sebastian Thrun

Abstract. Reliable vision-based grasping has proved elusive outside of controlled environments. One approach towards building more flexible and domain-independent robot grasping systems is to employ learning to adapt the robot's perceptual and motor system to the task. However, one pitfall in robot perceptual and motor learning is that the cost of gathering the learning set may be unacceptably high. Active learning algorithms address this shortcoming by intelligently selecting actions so as to decrease the number of examples necessary to achieve good performance and also avoid separate training and execution phases, leading to higher autonomy. We describe the IE-ID3 algorithm, which extends the Interval Estimation (IE) active learning approach from discrete to real-valued learning domains by combining IE with a classification tree learning algorithm (ID-3). We present a robot system which rapidly learns to select the grasp approach directions using IE-ID3 given simplified superquadric shape approximations of objects. Initial results on a small set of objects show that a robot with a laser scanner system can rapidly learn to pick up new objects, and simulation studies show the superiority of the active learning approach for a simulated grasping task using larger sets of objects. Extensions of the approach and future areas of research incorporating more sophisticated perceptual and action representation are discussed

Keywords: robots, vision, manipulation, active learning, grasping, interval estimation

1. Introduction

Vision-based grasping and retrieval of objects is an important skill in many tasks. A robotic system having the ability to perceive pertinent target object features, and the ability to select viable grasp approach and preshape values for a robotic hand would be able to carry out many useful functions. Possible scenarios for such as system would range from the handling of toxic materials in dangerous environments to the assistance of people with physical disabilities in relatively benign household environments. While significant progress has been made in the development of machine perception, highly dextrous robotics hands, and robotic grasp analysis and synthesis, limitations in the state of the art in vision and robotics still preclude the availability of "off-the-shelf" robot systems with truly effective vision and manipulation capabilities. This is primarily due to limitations in the reliability of machine perceptual systems in varied environments, as well as the difficulties in the control of dextrous manipulators. The environments into which a robot system is placed may have a variety of unfamiliar objects, materials and lighting conditions, making the perceptual and manipulation task difficult without the use of a significant number of domain-dependent techniques.

1.1. Previous Work

Previous work in robotic grasping may be divided into two major approaches: analytical approaches based on detailed theoretical analysis of the mechanics of contacts and heuristic approaches inspired by the research of neuropsychologists (Jeannerob & Biguer, 1981, Jeannerod, 1988) and orthopedic medicine (Schlesinger, 1916, Napier, 1956).

Analytical approaches to grasping have analyzed grasp selection based on mechanics, most often in terms of maximizing some objective function such as grasp stability, slip-resistance, or minimizing some objective function, such as internal forces (see(Cutkosky, 1989) for a survey). These approaches have given a great deal of theoretical insight into grasp planning, although they incorporate simplifying assumptions in order to be tractable and assume essentially perfect perception of relevant object properties such as shape and mass. Unfortunately, these *a-priori* assumptions may not be valid in all situations and may become less accurate as a system changes over time due to mechanical wear or other factors.

An alternative approach to selecting grasp approach and preshapes is to use a heuristic approach which employs domain-dependent rules based on objects and selections that are empirically observed. Heuristically-based grasp generators almost always include some equivalent of major grasp preshape types as categorized by Schlesinger (Schlesinger, 1916) including the fingertip, hook, cylindrical, palmar lateral and spherical grasp types. These different categories were formed based on functional analysis of human grasping behavior where different characteristics classes of hand shapes were observed to be associated with the prehension of different object shapes and tasks combinations.

A number of rule-based systems for grasp selection (Iberall, et al., 1988, Liu, Iberall & Bekey, 1989), were subsequently developed based on how skilled individuals chose grasps. These systems incorporate production rules that capture how humans choose particular preshapes and grasp sites. These approaches share a significant limitation with the analytic systems discussed so far in that they treat grasp selection as a completely separate process from perception, assuming perfect perception of object shape and other object properties. For these techniques to be practical, they require the machine perception system to provide descriptions at a level of detail and accuracy that is far beyond what can be provided by existing vision systems, especially if the variability of objects and the environment is taken into account.

One approach to solving the perceptual bottleneck is to retreat from general purpose systems and constrain the domain in which the autonomous system is to perform. Given sufficient constraints, a system may be engineered by taking advantage of *a priori* knowledge to simplify visual reconstruction of the scene and motor processing. Stansfield (1990) describes a heuristic rule-based system for grasp selection that uses a symbolic multi-component representation of object views derived from a range scanning system. If 2-D object contours can reliably be extracted from vision processing in a particular domain, then techniques employing analytical grasp point selection for optimal grasp on smooth closed 2-D contours can be used (Taylor, Blake & Cox, 1994). Alternative approaches such as Bard (1991, 1995, *in press*) compute ellipsoidal decompositions of voxel description that rely on fusing multiple stereo vision views in a space occupancy grid of voxels. The preshape generation process attempts to identify feasible ellipsoid parts for grasping according to heuristics based on part size and accessibility, followed by a physics-based simulation to verify a grasp before execution.

Unfortunately, retreating from a truly general purpose system by hand selecting reliable domain-dependent perceptual attributes (e.g. smooth object contours), feature extractors and special purpose grasping routines requires custom programming and engineering which can add significant cost to robot systems. Additionally, it is a great burden on the engineers and programmers to guarantee system reliability over a combinatorially large number of possible scenarios. Each time the system is deployed in a different domain, significant re-engineering effort may be necessary to adapt the system to function correctly. This additional cost makes it difficult to justify the use of robots as a truly flexible task solution. In future situations that may require true long-term autonomy, such as space exploration, there may not be engineering personnel available to carry out the required adaption.

1.2. Learning Approaches to Vision Based Grasping

One approach to avoiding the pitfalls of domain-dependent systems is to employ perceptual and/or motor learning techniques. Many researchers (Kuniyoshi, Inaba & Inoue, 1989, Kang & Ikeuchi, 1993, Dunn & Segen, 1988, Bennett, 1991, Kamon, Flash & Edelman, 1994, Salganicoff, 1992, Salganicoff, 1993) have taken the approach that it is preferable for robotic systems to learn a grasping strategy. The learning approach has the benefit that the system adapts to the characteristics of objects in its domain as well as to its perceptual and motor system, providing a system with higher autonomy. Robot learning systems can incorporate the true idiosyncrasies of their sensing and motor apparatus rather than rely on possibly inaccurate or outdated expectations of the system's designers. By taking a learning approach to vision-based robotic grasping (and robotic tasks in general), the noisiness and variability of both the perceptual, action systems and the environment can be incorporated into the robotic system to better ensure success, since many learning algorithms can handle noisy or missing inputs. Secondly, gradual drift or even catastrophic failure in sensor and motor systems and in the properties of the environment may be compensated for using learning without additional programming effort, leading to systems that are more robust and autonomous. If a sufficiently rich perceptual and motor representation is chosen, then the learning system can use these representations as needed over the many different environments where the robot might be placed.

Techniques in learning to grasp have taken both supervised and unsupervised approaches. Supervised approaches are exemplified by a rote learning or "teaching by showing" approached developed by Kuniyoshi et al. (1989) and Kang and Ikeuchi (1993). The observed history of actions is used to form a plan composed of action primitives that the robot can carry out. In general, these systems require significant *a priori* knowledge about the task and controlled environmental conditions in order to enhance the speed and reliability of the vision system in recognizing actions and objects in real-time.

Unsupervised learning approaches do not require human intervention and so are better suited for high autonomy. A number of systems have used the output of computer vision, primarily in the form of two dimensional edge contours, to drive either recognition and indexing of previously grasped objects, or to use the local object contour information as the basis of features for classifiers. Dunn & Segen (1988) employed an unsupervised memory-based learning approach with a vision derived two dimensional polygonal representation; grasp was coded by relative location and orientation to the object. During execution, a 2-D model matcher was used to index the presented object to a previously manipulated object and to invoke the previously successful grasp.

Kamon et al. (1994) used a contour-based two dimensional representation of the perimeter of objects and measured grasp location and human selected heuristic quality parameters for the object. During on-line learning the system tried to apply previously attempted grasps to the current object and then compared their predicted fitness using a nearest-neighbor learning rule. If no grasp had high predicted fitness, a randomized domain-specific heuristic was used to select a new grasp.

A knowledge intensive approach to learning to grasp was taken by Bennett (1991) who worked in robotic grasping of polygonal 2-D puzzle piece tasks using explanation-based learning and simple domain theories about uncertainty and grasping.

Tan (1990, 1993) employed a feature-based sonar depth representation and cost-sensitive learning (CSL) extension of ID-3 (Quinlan, 1986) to learn to recognize object labeled by high-level grasp-approach combinations (e.g. approach object from top gripping rim in pinch grasp).

Stansfield has also investigated replacing expert system production rules for grasp selection (Stansfield, 1990) with the batch-learning of associations using a connectionist representation with back-propagation learning (Rumelhart, Hinton & Williams, 1986). The system was trained to associate feature/view bindings to grasp categories and grasp sites.

Salganicoff (1992, 1993) developed a system for grasping objects based on full three-dimensional information from a laser range scanner and a parameterized superquadric representation. It learned to select from a predetermined set of canonical approach directions using a density-adaptive learning algorithm along with a forgetting component to allow learning set updates to incorporate changes in the task.

1.3. Motivation for Active Learning

The advantages of a learning approach to vision-based grasping, such as increased autonomy, and self adaptation to limits in sensing, action and the particulars of a given domain, are, of course, not without tradeoffs. In robot learning, as in many other learning tasks, the costs of acquiring the learning set may be the single greatest barrier towards the application of learning in a task. The acquisition costs for each exemplar can be particularly expensive, since each sensing and/or motor action often takes a significant amount of time to execute, and has other concomitant risks as well. In fact, the computational complexity of invoking the learning algorithm may be small in comparison to the cost in time and material of gathering the learning set (Tan, 1993). Therefore it is imperative to decrease learning set cost, and active learning provides an unsupervised way of doing so. Active learning systems allow the learner to control where in the input space their exemplars are drawn. They thus permit the learners to use strategies which balance the costs of gathering exemplars for learning (exploration) against the cost of misclassification during the execution of the task (exploitation).

Another way of looking at active learning is to view it as an optimization problem in which one is simultaneously building a model of a process and optimizing its performance. These problem has been studied by several different sets of researchers. In applied statistics, e.g. for the chemical process industries, evolutionary optimization (EVOP) has long been influential (Box & Draper, 1969). In EVOP, one fits a local linear or quadratic model to the available data, moves the process toward an optimum, fits the model again, moves again and iterates. These active learning methods are being applied today in large chemical plants (see (Moreno & Yunker, 1992)).

A second line of research comes from statistical decision theory and multi-armed bandits, and uses methods such as the Sequential Probability Ratio Test (SPRT), which rigorously computes whether one should collect further data points (see (Berger, 1985), and Gittins indices (Gittins, 1988)), which allow one to compare multiple arms (courses of actions) against a single reference arm, and thus avoid pairwise comparisons. As with all Bayesian methods, one must assume models of the distributions of the observations, including their correlation structures (if any) and one needs priors on the parameters in the models. The computations become exceedingly difficult for complex models of the probability distributions or costs.

A third line of research comes from the machine learning community, in particular workers in reinforcement learning and robot learning. Because of the unique demands of robot learning a number of active exploration approaches have been developed (Thrun & Moller, 1992, Moore, 1990, Schneider, 1993, Sutton, 1990, Atlas, et al, 1990, Cohn, 1994) to accelerate learning in those domains. Reinforcement learning researchers have also developed a variety of active exploration heuristics (Sutton, 1990, Kaelbling, 1990, Thrun & Moller, 1992) that tradeoff exploration and exploitation during adaptation.

This article presents a general technique and evaluation for active learning. The approach combines the Interval Estimation (IE) (Kaelbling, 1990) exploration heuristic with the ID-3 inductive learning algorithm (Quinlan, 1986) and does not require separate training and performance phases. Actions are then chosen using an indirect indexing scheme (Salganicoff & Bajcsy, 1992) which indexes the leaves on the ID-3 tree that have a high expected probability of success (reward) conditioned on the current perceived attributes.

We apply this framework to a simplified version of the vision-based grasping problem, where the approach axis for a two fingered grasp must be selected based on the extents of the three semi-major superquadric axes of the target object. The result is a learning system that smoothly combines its training and exploitation of learned knowledge, decreasing the number of failures encountered until good grasping rules are induced. This allows the learner to benefit from learning as it occurs instead of forcing the system to wait until the end of the learning phase to take advantage of the acquired knowledge. The general framework for the active learning algorithm IE-ID3 is described in section 2. The specific grasping problem instance, representation, experimental setup, and empirical and simulation results for the method are elaborated upon in section 3. In experimental trials with a robotic system with a set of 6 objects, the system learns to select reliable approach direction in a small number of attempts; convergence occurs in approximately 30 attempts per object. Simulations give a qualitative picture of the system's asymptotic performance as it interacts with a large number of objects. In simulations the IE-ID3 algorithm performs at almost twice the level of an open loop non-active learner, and also results in a higher classification accuracy. Section 4 compares IE-ID3 with other active learning and grasp learning approaches, and points out areas of future research, limitations and possible improvements. Finally, section 5 concludes with a synopsis of the approach and results.

2. An Active Learning Algorithm for Real-Valued Domains

Many approaches for action selection in active learning keep statistics relating to a particular discrete state or state/action pairs (Thrun, 1992). A number of different rationales have been used. For example, inversely weighing the raw number of times a state has been visited (Thrun, 1992) as a factor that favors execution of an action results in a system that is generally exploratory in nature, attempting to search the entire input space. Taking recency of a state visitation (Sutton, 1990) as a positive weighting for action selection gives a system which attempts to evenly visit the entire state space and track changes in the environment. These types of algorithms have been developed in the context of delayed reinforcement problems (Sutton, 1988, Watkins, 1989) with Markovian frameworks having discrete action and state framework, making them poorly suited for operation in continuous state and action spaces. However, it is possible to adapt techniques such as reinforcement learning to continuous domains through adaptive input-space partitions of those spaces. For example, Moore (Moore, 1991b) describes a modification to discrete state dynamic programming by the use of an adaptive resolution tree that partitions a continuous state-space into intervals. We employ a similar tree-based partitioning approach by using a classification tree algorithm (ID-3) (Quinlan, 1986, Quinlan, 1992) to form a partition over the continuous space, yielding an adaptively discretized space over which the Interval Estimation (IE) exploration algorithm ((Kaelbling, 1990), pp. 56), a discrete-space approach, can be applied.

The IE algorithm (Kaelbling, 1990) works by keeping statistics on the number of times a given action has been executed and the proportion of times that it has succeeded. A confidence interval is computed for the underlying probability of success for each of the feasible actions in a given context. The action whose upper-bound confidence interval is highest is chosen. The rationale is that an action may have a higher upper-bound for two reasons: either because the action is actually a good one to take, or because very few trials of the action have been executed, in which case there is insufficient empirical evidence to decide the goodness of the action in that context. Most importantly, this approach rapidly stops executing actions which have a very low likelihood of success and focuses on sampling those actions whose conditional probability of success cannot statistically be ruled out as best. If two actions share the same upper-bound, then they are chosen with equal probability.

We generalize the IE algorithm from its original formulation for a finite number of discrete actions to continuous valued perceptual and action spaces by combining it with the ID-3 algorithm (Quinlan, 1986) incorporating real-valued attribute splits. ID-3 automatically constructs decision trees by determining what questions (real-valued feature threshold values) give maximum information gain at each level of the tree. ID-3 and enhanced methods such as Classification and Regression Trees (CART) (Breiman, et al., 1984) and C4.5 (Quinlan, 1992) are widely used and can handle high-dimensional feature spaces, ignoring irrelevant features, and, when the trees are pruned to eliminate spurious branches, are relatively insensitive to noise in the data. The resulting ID-3 decision tree induces a hard partition on the real-valued attribute space with its leaves.

Consider an agent that perceives the world through a perceptual vector of continuous attributes P, and can continuously control the value of its action parameters through an action vector A, to yield a binary outcome O. The agent attempts to learn the mapping $O = f(P, A)$ over some sub-domain of possible P and A values. This function can then be used to maximize its reward by selecting actions A that maximize the probability of $O = 1$ given what is currently perceived in P. Therefore, rather than have an input space that consists solely of action attributes, we use an extended representation which consists of a combined perception-action space $(P, A) = ((p_1, \ldots, p_l), (a_1, \ldots, a_m))$, where p_i and a_i are the individual perception and action attributes for an agent.

At each step in the execution of the overall algorithm, we select an action as described below, execute it, ascertain its outcome, and then compute a classification tree T using ID-3 with real-valued splits, based on the distribution of binary outcomes O in the combined perception-action space. This cycle is repeated for the duration of the task.

2.1. *Indirect Prediction*

To select actions in the combined perception-action space, the system uses the tree T as an indirect predictor (Salganicoff, 1993) along with the IE method. First, the system senses the perceptual vector P. Let F be the set of leaves of the ID-3 tree that have perceptual interval values that intersect the currently sensed attribute values of P. The set F represents a partitioning or binning of the action space by the current decision tree subject to the perceived attributes, P, of the current object. As we vary A over all possible values with P fixed, we move over all possible leaves which predict outcomes for different action parameters A (see Figure 1). Indirect prediction is similar to the idea of using indirect control by taking the reward gradient of the reward function approximation with respect to the action space in order to hill climb to the action with maximum utility (Thrun, 1992). However in our case, the function representation is in terms of a tree decomposition, and is not differentiable, therefore we must do a partial-match search to maximize the probability of rewards over the feasible leaves of the tree.

Given this partitioning into bins in which different actions lead to different outcomes (all conditional on the perceptual vector P), we can directly apply the IE algorithm to choose among the possible actions. The statistics necessary for computing the confidence intervals (the number of success and failures contained in a leaf partition) are already available because they are needed to form the tree using the ID-3 algorithm.

For each leaf l_i in F we compute the upper-bound probability of success according to the binomial confidence interval formula (Larson & Marx, 1986). To calculate a probability interval that contains the true probability value with confidence $(1 - \alpha)$ given x successes (rewarding outcomes) out of n exemplars in the leaf so that

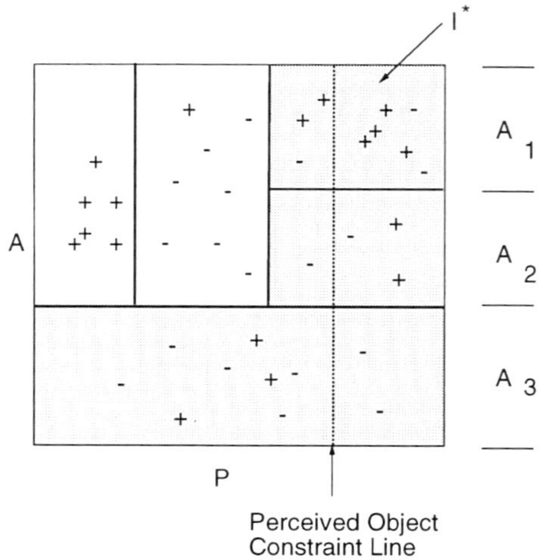

Figure 1. A schematic of the representation. The solid lines represent the ID-3 partitioning of the joint perception-action space (P, A). For the purposes of illustration assume a single degree-of freedom action A and a single sensed value P, although the approach generalizes to an arbitrary number of dimensions. When a new percept is sensed, this determines the discrete intervals for action values (the dotted line) and their associated statistics of task success (+) and failure (-). The feasible set of leaves F, is shaded. All feasible leaves are then compared in terms of the upper-bound confidence interval. An action is randomly selected from within the interval corresponding to the leaf with the highest upper-bound confidence interval.

$$P_{-} \leq P_{success} \leq P_{+} \tag{1}$$

where, P_{-} and P_{+} are the lower and upper-bounds of the probability interval estimate, respectively, one uses:

$$P_{\pm} = \frac{\frac{{z_{\frac{\alpha}{2}}}^2}{2n} + \frac{x}{n} \pm \frac{z_{\frac{\alpha}{2}}}{\sqrt{n}} \sqrt{\frac{{z_{\frac{\alpha}{2}}}^2}{4n} + \frac{x}{n^2}(1 - \frac{x}{n})}}{1 + \frac{{z_{\frac{\alpha}{2}}}^2}{n}} \tag{2}$$

where $z_{\frac{\alpha}{2}}$ is the confidence interval coefficient. The confidence interval coefficient is the deviation from 0 which is exceeded by a random variable with standard normal distribution with probability $\frac{\alpha}{2}$. Intuitively, the smaller α is chosen, the more exploratory the character of the learner, since more experiences are required to drive the confidence intervals down. In the limit, as $\alpha \rightarrow 0$, the system becomes completely exploratory and picks all actions as equally good, since the confidence interval upper-bounds approach 1. At the other extreme, as $\alpha \rightarrow 1$ the system become completely exploitational, since the upper-bound of the estimate approaches the raw empirical probability of the action succeeding.

The selected action leaf l^* is the leaf in F with highest upper-bound confidence interval as prescribed in the IE algorithm. Let A_{min}, A_{max} be the lower and upper action attribute intervals for l^*. The next action is chosen as random within l^* using a uniform distribution bounded by the intervals A_{min}, A_{max}.

3. Applying IE-ID3 to the Selection of Grasp Approach Directions

Robotic grasping and retrieval of unknown objects in unstructured environments is an important problem, and has many applications as discussed in the introduction. It is particularly suitable for learning-based approaches since the deployment of the system may occur in various environments. Many of those domains may be difficult to characterize ahead of time, making it difficult to guarantee the existence of specialized domain-specific perceptual and motor strategies that will succeed for all possible scenarios.

In our approach to vision-based grasp learning, we adopt the framework of the two phase model of grasping, consisting of ballistic approach and preshape phases as suggested by the work of (Jeannerod & Biguer, 1981, Jeannerod, 1988). Looking at learning as applied to the both phases leaves many free parameters to be set in both the ballistic approach phase that determines the direction of approach and finger contact locations, as well as in the preshape phase, where preshape category (e.g. pinch) is chosen and parameterized in terms of characteristics such as aperture, inter-finger span and finger stiffness. Similarly, from the perceptual standpoint, many other superquadric shape parameters and other sophisticated shape representations, including taper and bending, can be part of the perceptual representation. Because we had only a parallel jaw gripper available for the experimental setup we simplified our action representation to controlling only the approach direction of the gripper relative to the centroid of the object and held other action parameters fixed. We also simplified the perceptual representation to just the extents of the object's three semi-major axes, since we were most interested in exploring feasibility of the active learning approach using the simplest possible representation (see Figure 2).

This simplified representation was chosen for a number of reasons. Our previous experience with our grasping system showed that running experiments with the our laboratory setup up was a frustrating process, due to the unreliability that is typical of laboratory prototype setups. This frustration motivated us to pick minimal representations of perception and action to see if our active learning approach could succeed with a very small number of grasping attempts during a single session of grasping. It is important to note that this does not imply a fundamental limitation of the learning system we developed, since the decision tree learning algorithms used are designed for higher dimensional spaces and can easily accommodate the addition of richer descriptions of object shape, both for categorical and real-valued attributes (this is further discussed in section 4).

3.1. Experimental System

The robotic system which carried out the experiments consists of two PUMA 560 robots: the perceptual robot and the grasper. The pertinent aspects of the system are summarized in Figure 3. The range image provided by the scanner (see Figure 4) is processed using a superquadric data reduction procedure (Solina & Bajcsy, 1990) which fits a superquadric to the dense range data. The superquadric is a necessary data reduction step, as the range image can contain thousands of points. It was chosen as a representation since it yields a compact set of parameters that attempt to approximate the overall geometry of the object to be manipulated; implications of this perceptual representation are addressed in section 4.

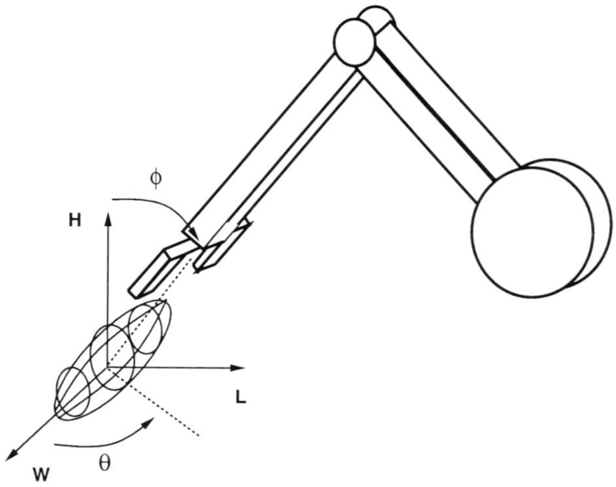

Figure 2. The perceptual and action representation for the grasping system.

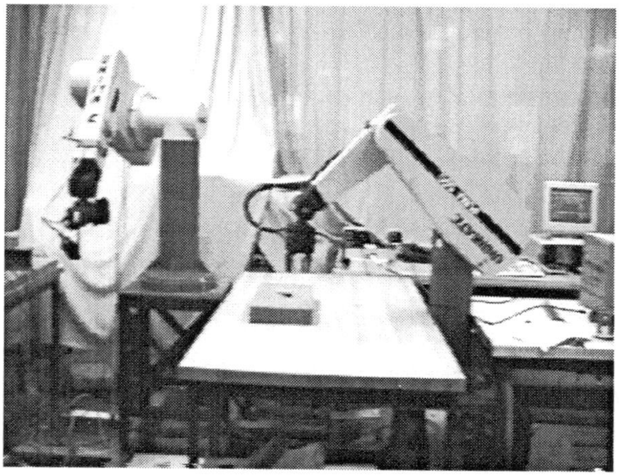

Figure 3. The experimental system, (a) consisting of the perceptual robot (left) and the grasping robot (right). The perceptual robot first scans the unknown object to yield a range image. A set of parameters is found which best fits the range data and this is used as a perception input to the learning system. The grasper then attempts to pick up the object based on the percept plus its acquired knowledge about the effectiveness of different grasp approach directions.

Figure 4. The vision system consists of a mobile laser range scanner mounted on the end of the perceptual robot which is swept over the object to be grasped. This yields a sequence of height values which can be combined into a complete range image. This range image is then processed by fitting a parameterized surface (a superquadric) to the height values in the range image. The resulting parameters for the surface are then the input attributes to the learning system along with the chosen action.

Our Laser Range Imaging System consists of two components: The *LOOKER* and the *GUS* processing unit (see Figure 4). The *LOOKER* is composed of a laser stripe generator and SONY XC-39 camera which generates video signal of the images obtained under the illumination of the laser stripe, and the *GUS* unit (Tsikos, 1991) processes the continuous sequence of laser images and generates a static range image of the scene.

The range image is first cleaned by morphologically eroding it in order to remove any spurious outlying measurements and then sub-sampled (see Figures 5 (a,b)). The resulting range points then form the constraints for the numerical optimization to extract the parameters for a surface which best-fits the range points. This yields the superquadric transformation and shape parameters in the image coordinates (see Figure 5 (c)). These values are then transformed into the scanner robot task frame by using the scanner-to-world transformation matrix. The resulting superquadric is put into canonical form, with $a_x \leq a_y \leq a_z$ and with the a_x, a_y and a_z values bound to h, w and l depending on the pose of the object to gravity. It is assumed that the object is placed on a flat surface perpendicular to the gravity vector. An object-centered action frame is chosen so that the θ and ϕ are relative to the direction of the w and h major axes and aimed at the visual centroid of the object.

The success or failure of a grasp is assessed by using the force-torque sensors on the gripper. During the grasp attempt, the system moves the gripper fingers inward from their full open position to an inter-finger distance of 5mm (see Figure 6). After the approach and grasp phase, the robot moves the gripper upwards, hopefully lifting the held object. It then measures the normal forces on the gripper fingers. If the force is below a threshold value, then the fingers are characterized as empty, otherwise if there is a large normal force, the grasp is labeled a success.

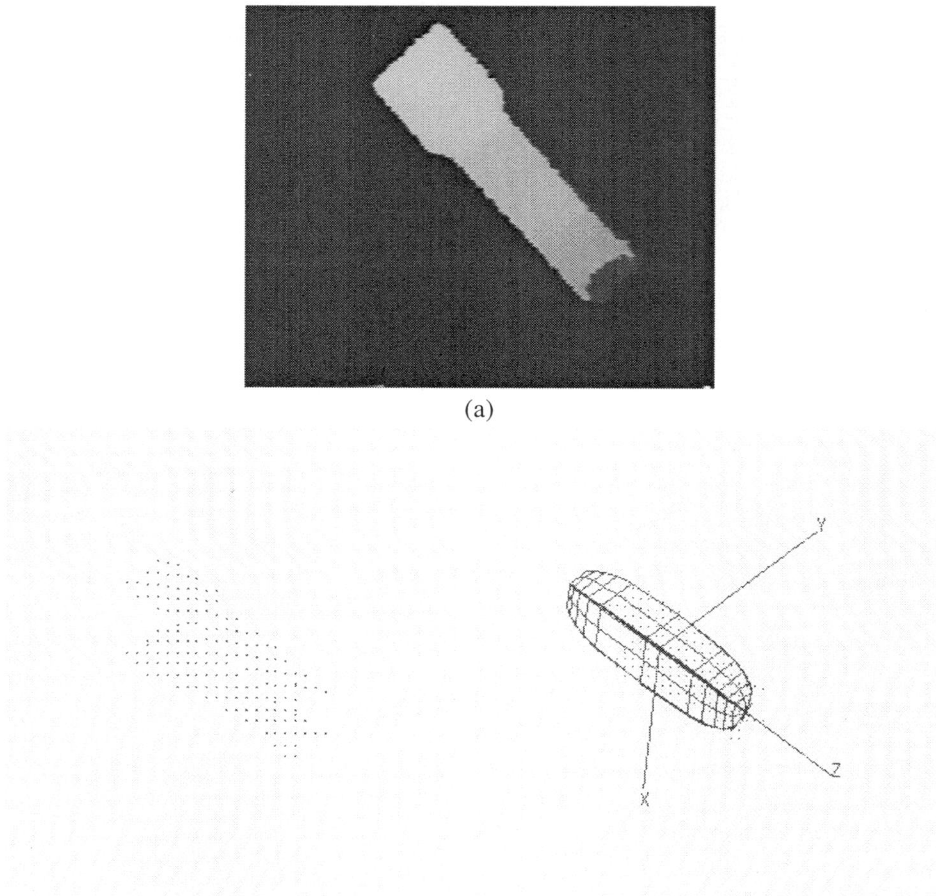

(a)

(b) (c)

Figure 5. Acquired 3 dimensional Range Points (a) for the stapler, the sub-sampled and eroded points (b) and the resulting Superquadric fit to the range points (c).

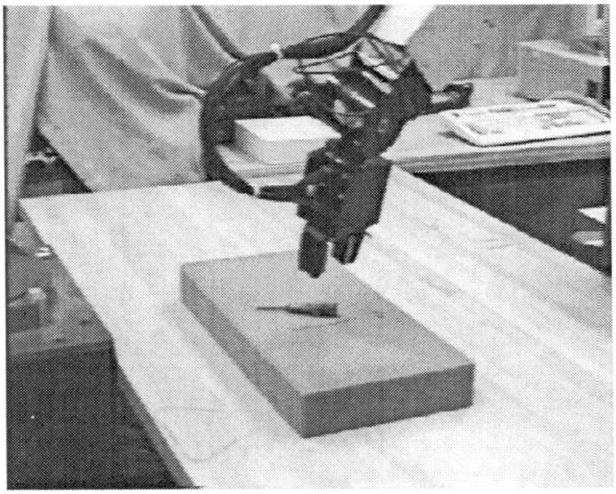

Figure 6. The gripper is a simple two finger gripper with force torque sensors to sense grasp force. During the grasp attempt the hand attempts to close the fingers to their minimum aperture. The finger motion is guarded, so that if an above threshold force is detected, the motion stops. After a grasp attempt, the hand is brought upward and the force sensors are used to measure whether the object has been successfully grasped by again monitoring the reaction forces on the fingertips. Compliant foam is placed under the object to prevent damage to the gripper or object in situations where the fingers would be stubbed and possibly damage the object or gripper.

3.2. *Experimental Protocol*

In summary, the operation of the system consists of the following sequence:

1. The laser scanner scans the object.
2. Data reduction algorithms refine the data and produce the h, w and l of attributes that describe the object (Solina & Bajcsy, 1990).
3. A full decision tree is regenerated based on the database of points using the ID-3 algorithm for real-valued attributes (Quinlan, 1986, Quinlan, 1992) [1].
4. The tree is pruned ($\alpha = 0.025$, see below). In practice, the tree pruning parameter did not seem to have a significant impact on system performance for reasonable values.
5. Interval estimation is used (with confidence level $\alpha = 0.125$) to select the best θ, ϕ using the approach described in section 2. Here $P = (h, w, l)$ and $A = (\theta, \phi)$.
6. The robot attempts the grasp and assesses the outcome.
7. The new data point (P, A, O), of attributes and the outcome, is recorded in the database.
8. Goto step 1.

Pruning is done using Quinlan's (Quinlan, 1992) C4.5 pessimistic binomial upper-bound method. For each leaf the $(1 - \alpha)$ upper-bound on the estimated error rate is computed

Figure 7. The set of common objects for which learning was attempted.

based on the misclassification incidences for each of the children nodes below a node being evaluated for collapse. If the weighted average of the upper-bound error rates is greater than the pooled upper-bound error rate, then the child nodes are collapsed.

3.3. Experimental Results

A total of six objects were chosen for grasping: a balsa block, white-board eraser, small flashlight, wire spool, screwdriver and a stapler (see Figure 7). For each of these objects, the protocol of section 3.2 was carried out, starting with an empty learning set for each new item.

Figure 8 shows the average success (or hit) rate over the thirty trials for the balsa block. The hit rate is the windowed proportion of successful grasps to attempts using a sliding window 10 attempts wide over the 30 instances in a learning run. For example, the computed hit-rate for trial 10 consists of the proportion of consecutive grasp attempts numbered 10 through 19 that succeeded. It can be seen that the robot rapidly converges on a strategy for selecting approaches that succeeds reliably.

The scatter plot of Figure 8 also illustrates the effect of the exploration strategy. This figure shows the selected azimuth θ and elevation ϕ values for the balsa block trials in the range $0° < \theta < 90°$ and $0° < \phi < 60°$. Each point in the θ, ϕ space is marked by its outcome, 0 for failure or 1 for success. As is evident by the plot, the action parameters are not uniformly distributed in the space, but concentrated in the region which has a small θ value, which is the constraint for alignment for this object. Outside of the constraint values there are very few grasping attempts. The exploration algorithm has rapidly focused exploration on the region of success.

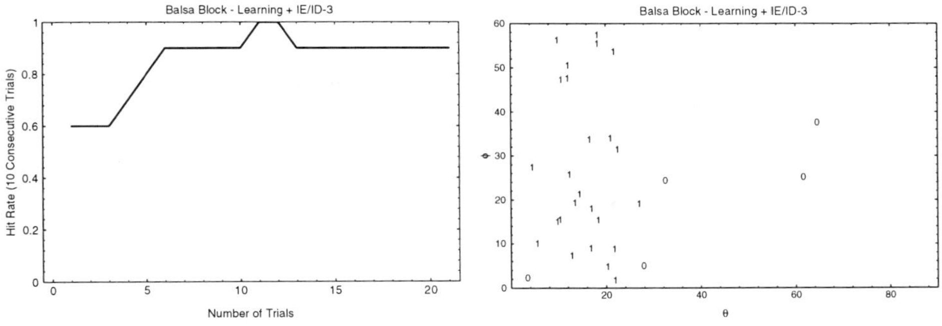

(a) (b)

Figure 8. The hit rate for the balsa block over 30 trials and (b) corresponding scatter plot of chosen actions in the action space. The cumulative probability for successful grasping is .83 for the learning condition and .43 for random approach.

In the case of the flashlight, the learning curve increases more slowly (see Figure 9) since the width of the screwdriver is quite close to the maximum width of the gripper and it takes more attempts to find a first successful grasp which can form the basis of a tree partition.

The learning curve for the stapler is rapid (see Figure 10), converging after about 20 attempts. The system succeeds almost at once and maintains a high level of performance. The scatter plot once again shows that the actions are highly concentrated in the favorable part of the action space. Similar learning performance and action distributions were observed for the eraser and screwdriver objects.

The spool (rightmost object in Figure 7) was an interesting case, since it has a significant concavity from the end-pieces attached to the center shaft, which makes it difficult to represent accurately using the convex superquadric representation that we have chosen. Nevertheless, the learning algorithm is able to learn the constraints for grasping this object (see Figure 11), and identifies a region in the action space that succeeds (namely $\phi <$ 30 degrees) by picking an approach direction which avoids approaches that hit the end-cap, which was the main cause of failures in grasping this object. Unfortunately, the system cannot describe concavity with its given shape representation. This may lead to the generalization that other objects with similar parameter values have constraints for approach directions that are overly restrictive when compared to a truly convex object. Another scenario has a convex object with the same extents as the spool presented first in sequence. This will lead to approach constraints which are not restrictive enough, and upon interactions with the spool, the system may err by selecting poor approach directions. Upon further interactions, the system will begin to restrict the approach directions. Unless the representation is extended to include some measure of concavity, this problem cannot be avoided. The only solution is to adopt more sophisticated representation for perceptual parameters into the active learning approach (see section 4).

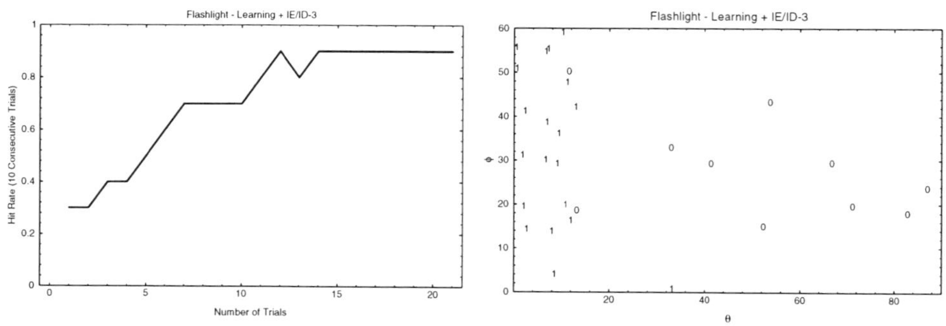

(a) (b)

Figure 9. The hit rate for the flashlight over the first 30 trials and (b) the distribution of actions in the action space. The cumulative probability for successful grasping is .67 for the learning condition and .23 for random approach.

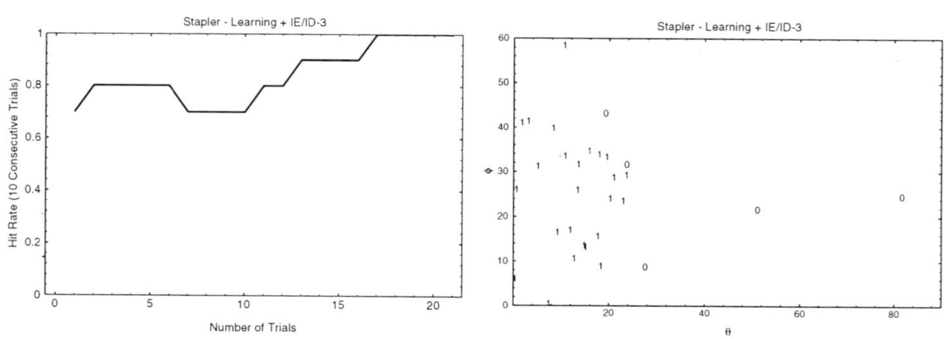

(a) (b)

Figure 10. The hit rate for the stapler over the first 30 trials and (b) the distribution of actions in the action space. The cumulative probability for successful grasping is .83 for the learning condition and .4 for random approach.

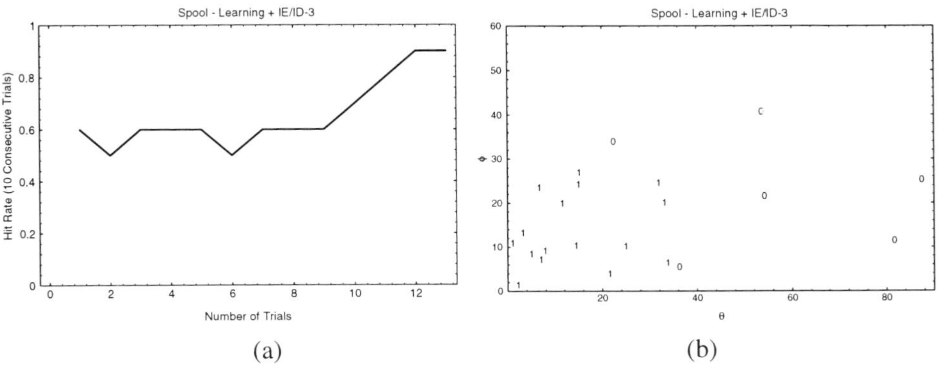

Figure 11. The hit rate for the spool over the first 25 trials and the distribution of actions in the action space. The cumulative success probability is .8 for the IE-ID3 learning condition and .27 for the random condition.

3.4. *Simulation Studies*

A simulation study was undertaken to explore the scaling properties as the number of objects presented increased. In the simulation the gripper was modeled by taking the necessary coordinate transformations on the gripper of rotation(θ, z) followed by rotation(ϕ, x) and determining if the projection of the object in the two dimensional plane orthogonal to the approach direction was smaller than the maximum span of the fingers. If the gripper cleared the object, it was taken to be graspable.

The objects for the simulation had dimensions that were selected from a uniform distribution of 35-90mm and the maximum gripper width was taken as 70mm. Additionally, to make the task more difficult, objects were generated so that only one of the aspects of the object, w or l would clear the gripper. For the aspect that did clear the gripper, a minimum of 5mm clearance was guaranteed. The allowable approach directions were in the range $0° < \theta < 90°$ and $0° < \phi < 90°$.

For the IE-ID3 algorithm, the learning protocol was to give a randomly generated object from the above distribution to the learner and then have it attempt to grasp the object 10 times, or until it succeeded 5 times in a row, whichever came first. Each trial consisted of running ID-3 on the exemplars generated so far and then selecting an action using the IE algorithm on the resulting tree. For the simulation curves, the horizontal axis represents the interaction number as they occur in each independent execution epoch, and the vertical axis represents the estimated hit rate (the probability of a grasp attempt succeeding) based on the average success rate at each trail interaction in the epoch. The estimated hit rate (success) is computed by counting up all successful attempts for the given trial numbers across all epochs and dividing by the total number of epoch runs.

For comparison, open loop (OL) alternating learning was performed in which each cycle consisted of first executing a random action to acquire the next point in the exploration set. Then ID-3 was run on that exemplar plus any previously acquired exemplars and a test action was executed to form a trial for the performance curve. The test action was chosen by

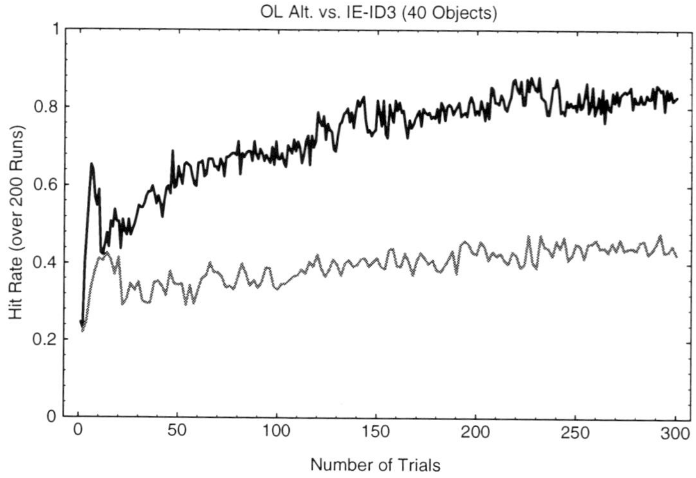

Figure 12. Comparison of system performance over first 300 trials (averaged over 200 runs). The IE-ID3 driven learner (black) exhibits much better performance than the open loop learner (gray) which suffers because it always alternates between the random training action mode and test.

searching for the leaf with the highest raw probability of success given the current perceptual vector constraint P. The OL plot depicts the average of the exploration and exploitation outcomes. As Figure 12 shows, the IE-ID3 based learner (the black curve) performed almost twice as well the OL alternating strategy. This large difference in performance was primarily due to the fact the in the OL case 50% of the grasp attempts were randomly chosen because of the random selection used in OL.

Figure 13 depicts the cumulative average number of successes for three conditions of learning: IE-ID3, OL and random. Again we can see that the performance of the IE-ID3 mode learning far exceeds the alternating OL explore/exploit paradigm again because of the wasteful random search of the open loop learner. This can be seen in Figure 14 which shows the performance curve (black) and exploration curve (light grey). Since the performance on the exploration is low, the combined performance of the alternating explore/exploit strategy is quite low; as expected, random exploration is expensive. IE-ID3 learning performs much better since it begins to exploit knowledge and abandon exploration as soon as is judged possible. Subsequent actions are guided to focus on more promising parts of the action space. Figures 15(a) and (b) give representative projections of the distribution of actions for the IE-ID3 and OL conditions along the θ and l axes and show that, as in the actual experiments, the data is concentrated in the high success regions for the IE-ID3 condition, as expected.

We would expect that along with better on-line performance, the classification accuracy for the active learner would be better than the open loop, since it would concentrate its data gathering in relevant regions of the input space, leading to better approximations of task-relevant concept decision boundaries by the learner. In order to test this hypothesis,

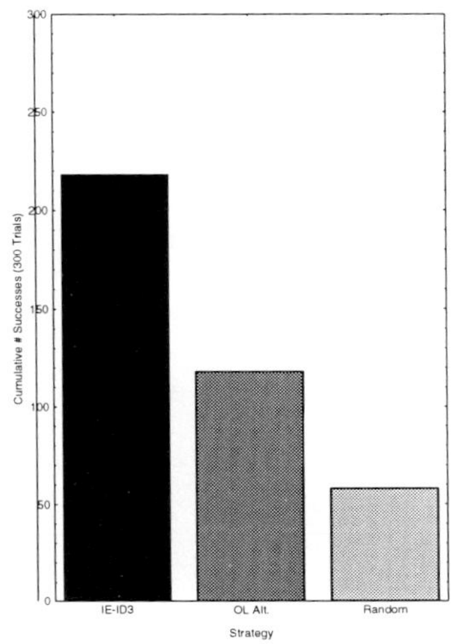

Figure 13. Comparison of the cumulative number of grasp successes for the IE-ID3, alternating explore and exploit open loop (OL) and random conditions over 300 trials.

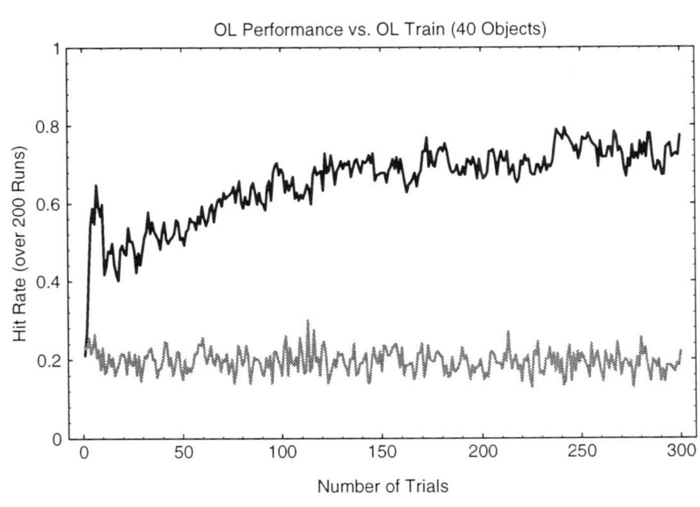

Figure 14. The exploration and exploitation performance of the open loop learner. Since these two strategies alternate, the combined performance is penalized by the random exploitation.

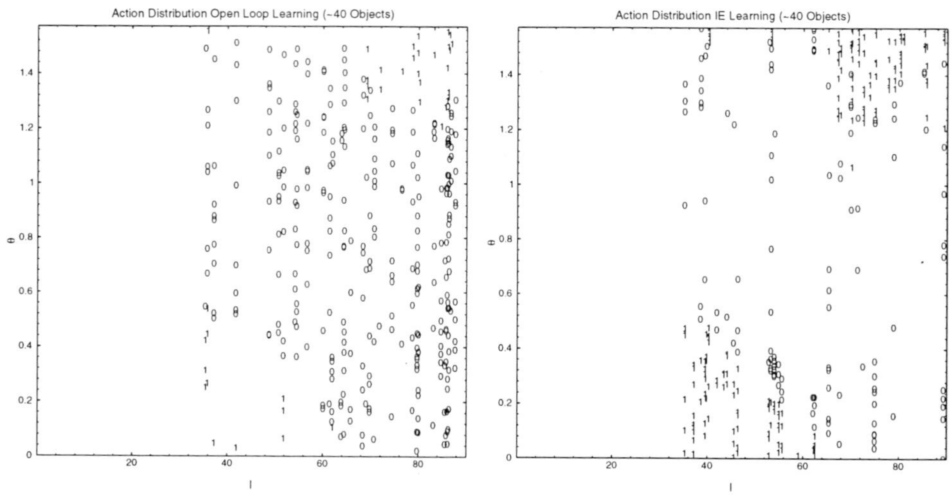

(a) (b)

Figure 15. The projection of the location of actions (θ) vs. the length aspect of the object (l). We can see that the exemplars generated by the active learning case (b) are much more concentrated in the success regions than the open loop case (a).

we ran the OL strategy and measured the probability of successful classification during the performance cases of each cycle. We also formed an equivalent independent test set for IE. Although a test set is technically unnecessary while using the IE-ID3 method, for the purposes of comparison it is useful to examine the use of raw probabilities for action selection of both methods. Figure 16 shows that the IE-ID3 formed classification is more accurate. In another test, IE-ID3 versus OL were run over 250 interactions using the same protocol as above. Looking at the average number of correct classifications for the 250 interactions for IE-ID3 ($\mu_{OL} = 151.121$ and $\mu_{IE} = 175.12$) showed that IE-ID3 lead to significantly higher number of correct classifications. The 99% confidence interval for the difference between the means $\mu_{IE} - \mu_{OL}$ was $[18.35, 29.4657]$ using a Student's-t small sample confidence interval test. Thus, the simulation results indicate that IE-ID3 does somewhat better than OL from a classification standpoint, but is markedly better from a *performance* standpoint, which is the key issue for real-world efficiency.

4. Discussion

We adopted the framework of the two phase model of grasping, and used learning to determine free parameters for the ballistic approach phase that determines the direction of approach of the robot hand. Our results show that the IE algorithm provides an efficient mechanism for focusing the data-gathering activity of an ID-3-based learner on successful areas of the action space. The method learns to grasp new objects very rapidly, usually within about 20 attempts for this simplified problem. The scatter plots for both the real

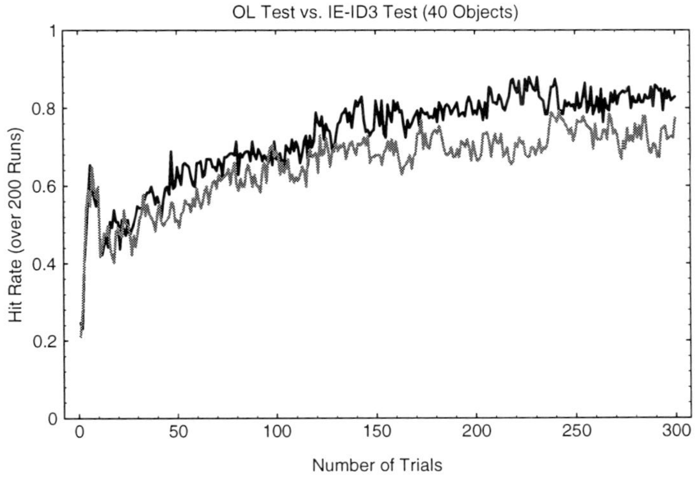

Figure 16. Comparison of the Classification Rate for IE-ID3 (black) Test vs OL. Test. (Grey)

learning trials and the simulations illustrate the fact that actions are concentrated in the high success areas of the attribute space.

4.1. Extending the Approach to More Complex Objects and Effectors

Our work represents a first step, where we are controlling only the approach direction of the gripper relative to the centroid of the object, and holding other parameters fixed. Our active learning approach to grasping performed well with a small parameter set, but further trials with richer perceptual descriptions are needed to verify to the approach with the more complex representation needed to represent general real world objects. In particular as more complex objects with multiple parts and other protuberances are presented for grasping, a global deformation approach such as the superquadric will smooth out these possibly important features (see the discussion of performance on the spool test object in section 3.3). Objects that might be graspable from handles, but too wide to clear the gripper might be found ungraspable using the global deformation approach of superquadrics, leading to decreased performance.

Fortunately other superquadric shape parameters and more sophisticated shape representations including taper and bending parameters for superquadrics could also be used within this framework. In particular, the multi-component aspect polyhedron described used by Stansfield (Stansfield, 1990) seems a particularly suitable candidate. This representation supports symbolic 3-D attributes of object patches such as curved or flatness properties, as well as 3-D positions and extents in space and other predicates that express the spatial relations between patches for a given view. These views from different directions of the object can be combined to form an aspect polyhedron. The aspect polyhedron permits the representation of objects with multiple parts, rather than the single global deformations permitted by superquadrics. It allows a more sophisticated representation of both of perception and action, since grasp sites can be associated with segmented surface patches on the object. In order to apply learning to this representation, the different slots of the multi-part perceptual representation could be bound to the input attributes of the decision tree (similar to the input bindings to the input nodes in (Stansfield, 1992)) but using the same protocol and active learning approach described in this work instead of an off-line back-propagation approach. A similar approach could be taken with Bard's hierarchical ellipsoid representation (Bard, Troccaz & Vercelli, 1991), if it were augmented with spatial-relation predicates between the component ellipsoids.

The IE-ID3 approach should scale well if these more sophisticated representations are adopted since all of its components are known to scale well. The ID-3/CART family of classification tree algorithms have proven effective in applications with much higher dimensionalities than what we have used here, and support both real-valued and categorical variables, both of which may be present in grasping tasks. Future implementations should include symbolic values, such as the grasp preshape (e.g. pinch, hook, cylindrical, spherical etc.) for more complex dextrous hands, along with real-valued parameters for those grasp categories such as finger stiffness and extent of preshape. Additionally, ID-3 and related techniques such as CART and newer non-axis parallel techniques (Murthy, et al., 1993) are very well-suited to rejecting irrelevant attributes because of the information gain splitting heuristic utilized, and so should scale well to more complex representations. Another advantage of using classification trees is that although the action space is continuous, using the indirect indexing method with the ID-3 partitioning affords a natural and meaningful partition of that space. The resulting intervals provide a range of feasible action which can be intersected with other high-level constraints, such as approach limits that might be due to obstacles around the object.

Further phases of grasping that follow the preshape and approach would also benefit from active learning. These include the contact phase and lifting phases (Howe, et al., 1990), re-grasping, transport phases and release or insertion for assembly (Gullapalli, Barto & Grupen, 1994), each of which is indicated by specific sensory markers, such as tactile forces, that invoke differing control modes. An interesting problem is to learn reliable markers from this stream of sensory information that would indicate state transitions, as well as the correct control parameters that would hold during these states, as has been demonstrated is some initial work by (Pook & Ballard, 1993) using hidden Markov models.

One possible drawback of the approach is that it is possible that other potentially successful areas of the perception/action space might be missed, since the method could greedily identify a good region and cease exploration in other regions. For example, an initial split based on the trials seen so far might yield a partitioning which covers only a subset of the true success region. Subsequent attempts would then be focused in that restricted region. There is no way to guarantee not missing a small region with higher success probability unless one is willing to exhaustively explore the entire perception-action space. Generally, the space is sparse in success for non-trivial problems, and it is not worth the cost of doing so.

Another possibility is that the IE algorithm may get stuck in actions which are sub-optimal due to pathological data orderings (Kaelbling, 1990). Consider the case two actions, a_1 and a_2, with $P_{success}(a_1) > P_{success}(a_2)$. When sampling a_1 and a_2 sequentially using the IE strategy it is possible that a pathological ordering of outcomes may occur leading to $P_+(a_2) > P_+(a_1)$ and $P_+(a_1) < P_{success}(a_2)$ at the same time. In this case the algorithm will be stuck in a vicious cycle: a_2 will alway be selected, since its upper-bound will converge to $P_{success}(a_2)$ and a_1 upper-bound will never rise since it is never selected since a_2 has the best upper-bound. Sticking may also occur in Bayesian-optimal systems; it is an unavoidable consequence of the exploration/exploitation tradeoff. Fortunately, sticking becomes less probable as the difference in the underlying probability of success between actions increases. Therefore, for practical purposes, if the sticking phenomenon occurs, it is likely that the efficacies of the different alternatives will be small.

4.2. Comparison to Other Methods

Active learning approaches have taken two overall approaches. Knowledge-directed approaches select actions which maximize overall knowledge gain about a domain model, independent of a particular task. Task-directed approaches concentrate on actions that yield a maximum of information about the domain that is relevant to a particular task in a domain, rather than a general domain model. For example, Cohn has proposed two methods for control of experimentation to maximize overall information gain. Selective sampling (Atlas, et al, 1990) attempts to identify the region of uncertainty between the most general and most specific concepts that are consistent with the training set. The distribution of query points is then biased to favor those points that lie within the region of uncertainty. Actions can also be chosen to be maximally informative in terms of minimizing the expected generalization error over the entire domain, for example, using optimal experimental design (Cohn, 1994). Whether a task-directed or knowledge-gain directed gain approach is taken for learning is a function of the costs of experiments and the expected amount of the domain that will be visited by the learning agent during its lifetime, and how desirable it is to front-load learning of the entire domain, versus learning it on an as needed basis as new tasks are encountered.

The system presented is the first grasp learning system to incorporate active learning to intelligently select manipulatory actions. Previously developed grasp learning systems adopt random action search, or rely of domain specific heuristics. For example in Dunn & Segen (1988) if no similar object is found in the data base, a randomized trial and error search for successful grasps is undertaken. In Kamon et al. (1994) if no match is found, a domain dependent heuristic is used to pick a grasp with a random component. Similarly in Tan (1993), although consideration is given towards picking good sensing actions, manipulatory actions are chosen randomly if no classification is predicted. Bennett's explanation based system (Bennett, 1991) uses domain knowledge to choose its next grasping attempt, which requires explicit human provided knowledge regarding the general characteristics of the sensing and motor apparatus of the robot, limiting its flexibility and true autonomy. Our approach represents a domain-independent and principled technique for selecting actions.

Our work can be compared Tan's Cost Sensitive Learning (CSL-ID3) (Tan, 1993) at several levels. Both CS-ID3 and IE-ID3 are similarly concerned with decreasing the cost of learning, but differ philosophically on how they decrease this cost. If one examines the overall costs encountered by a learning system during its lifetime, these costs can arise either from measuring the attributes of each case processed, or due to errors that occur by acting on the wrong classification output by the learning system (executing the wrong action). Consider a hypothetical system that attempts to learn to grasp and retrieve Ming Dynasty vases. Certainly, we would want the system to converge as quickly as possible, since each dropped vase has very high penalty. Secondarily, if certain perceptual features, such as the color of the vase, were rapidly computable and relevant to deciding where to grasp the vase, we should prefer those over difficult to sense and unreliable object features. If the cost of failure is high in comparison to sensing costs, then we want the system improve its classification accuracy on encountered cases as rapidly as possible. This can be done by intelligent selection of the next action, as in IE-ID3 and other approaches (Thrun & Moller, 1992, Moore, 1990, Schneider, 1993, Sutton, 1990, Atlas, et al, 1990, Cohn, 1994) so as to improve the classification accuracy as much as possible per each fixed cost exemplar. Until it reaches convergence CS-ID3 seems to require many more interactions with the object than a standard ID-3 and encounters many more classification errors than a standard ID-3 approach ((Tan, 1993),pp. 27), which can be problematic in tasks with high misclassification costs. On the other hand, if misclassification costs are small in comparison to sensing costs, the CSL-ID3 approach of decreasing average sensing costs makes sense. In fact, CS-ID3 achieves rough parity with ID-3 regarding total cost until convergence by decreasing the average sensing cost per each case.

Although initially developed as a supervised learning system (Tan, 1990) CSL-ID3 can technically be run in either a supervised or unsupervised mode (Tan, 1993), where either a teacher suggests the appropriate grasp for a sensed object, or the robot randomly tries grasps from a finite set of available alternative grasping procedure until a grasp succeeds. However if size of the action set is large, then the random experimentation approach can be unduly expensive and even hopeless. Secondly, if an object is not graspable by any of the preset grasp actions in CS-ID3, the system cannot synthesize a new action that succeeds. The IE-ID3 approach, on the other hand, allows for continuous actions, which do permit truly novel actions to be discovered. Secondly, because of the probabilistic nature of the confidence intervals we use, we can have a measure of reliability of an action, which is not captured the recognition tree generated by CSL-ID3

Since CSL-ID3 achieves a decrease in the on-going sensing costs of the learning agent after it has converged, it can lead to much lower costs during the lifetime of the system. To truly minimize costs during the life of the agent, the sensing and misclassification costs, as well as the expected number of times the gathered knowledge will be used (e.g. the lifetime of the agent) must be taken into account. An interesting avenue of future research would be to combine a technique such as CS-ID3 with IE-ID3 in a multi-objective approach, taking into account both of the above considerations.

4.3. Extending the IE-ID3 Approach to Other Learning Algorithms

The key to the approach described here is the combination of learning mappings from perception and action to reward with selecting actions which improve the expected reward over all future trials. Above, we used ID-3 to build decision tree mappings and the IE algorithm to select actions. Other learning methods such as radial basis functions, nearest-neighbor approaches and hierarchically gated mixtures of Gaussians could equally well be used to build the mappings since they also can, in principle, provide confidence intervals for the quantities they predict.

One immediate extension would be to apply the IE-ID3 approach toward learning function with continuous outputs. This could be done by employing the real-valued extension to IE as suggested in (Kaelbling, 1990), (pages 74-75), along with using regression trees (Breiman, et al., 1984) in place of classification trees. The real-valued extension for IE-ID3 involves building confidence intervals for the estimate of the true mean reward for the learner, again picking actions with the highest output upper-bound. The statistics for the upper-bound would be available in the nodes of the regression trees.

Additionally, one can use exploration approaches other than IE, for example, those suggested by statistical decision theory (Berger, 1985). These can incorporate priors on the distribution of reward in the attribute space in order to control exploration in a less heuristic fashion than IE, and allow a more formal analysis of the properties of the IE approach. Also, a wealth of results from multi-armed bandit problems (Gittins, 1988) become applicable once the real-valued space is partitioned and can lead to even more efficient solutions to problems (Salganicoff & Ungar, 1995).

4.4. Other Application Areas

We expect the IE-ID3 method to be broadly applicable in robotics and in many other domains. A variety of other robotic skills have been subjected to learning approaches through parameterization of the perceptual and action spaces, such as throwing (Schneider, 1993), juggling (Aboaf, Atkeson & Reinkensmeyer, 1988) and playing billiards (Moore, 1991a) and the IE-ID3 active learning approach should be applicable those domains as well.

Chemical plant control and optimization provides a second example of a class of problems with the right characteristics for such active learning, particularly IE-ID3. The plants are often nonlinear, and complete, accurate, models are rarely available. Many control actions can be taken - e.g. adjusting the temperatures, pressures, and flow rates of the plant. It is far too expensive to explore the whole range of possible actions (controls), since most of them are suboptimal - and expensive. Thus, it is necessary to trade off exploration of actions against exploitation of the best actions found to date. Similarly, in many batch production environments, such as semiconductor processing, it is difficult or overly expensive to build detailed and accurate models of the production process that may permit a principled enhancement of yield as a function of the variation of the process parameters. By using the IE-ID3 active learning approach an efficient search of the parameter space of the process may be made, even for environments with combinations of real-valued and categorical parameters.

5. Conclusion

We have presented a technique for active learning which combines the Interval Estimation (IE) exploration heuristic (Kaelbling, 1990) with a classification tree algorithm, ID-3 (Quinlan, 1986). We demonstrated the success of this technique in selecting parameters for the grasp approach direction during the ballistic phase of grasping. The active learning approach was motivated by the fact that gathering each exemplar (interaction with an object to attempt a grasp) has significant costs, similar to many learning tasks. Thus, minimizing the number of trials before good performance is attained is crucial. The resulting learner avoided random exploration and focused its data acquisition areas with effective actions. Empirical tests on a small set of test objects using a laser ranger scanner and simplified superquadric shape descriptions of the objects showed that the system was able to learn to select successful approach direction rules in a small number of real interactions with the objects. Simulation results for a scenario with many more objects and trials showed that using the active learning exploration strategy dramatically improved performance over an uninformed batch learner. The representation and learning algorithms employed are extendable to more sophisticated perceptual and action representations for grasping. Although the approach is promising, much work remains to be done in verifying the scalability of the approach through experiments and in tackling learning as it applies to other aspects of vision-based robotics and grasping.

Acknowledgments

This work was supported by Navy Grant N00014-92-J-1647, AFOSR Grants 88-0244, AFOSR 88-0296; Army/DAAL 03-89-C-0031PRI; NSF Grants CISE/CDA 88-22719, IRI 89-06770, and ASC 91 0813 and an NSF HPPC Postdoctoral Associateship for MS (CDA-9211136) at the University of Pennsylvania. Applied Science and Engineering Laboratories funding was provided through the Rehabilitation Engineering Research Center on Rehabilitation Robotics– National Institute on Disabilities and Rehabilitation Grant #H1333E30013 and Nemours Research Programs. We also thank Len Kunin for his help in implementation of the system.

Notes

1. The ID-3 algorithm with single partitions per attribute has worst case time complexity of $O(n^{2+\log k})$ (this solves the recurrence relation $T_n = 2T_{n/2} + kn^2$) where n is the number of exemplars, and k is the number of attributes. In practice performance is much better, since the above analysis assumes a learning set that results in a full binary tree partitioning of one leaf for each exemplar in the set. Truly incremental versions of ID-3 such as ID-5 (Utgoff, 1988) could have been used, but the non-incremental algorithm executed rapidly enough that it was not a hindrance to regenerate the full tree after each new observation.

References

Aboaf, E.W., Atkeson, C.G. & Reinkensmeyer, D.J. (1988). Task-level robot learning. In *Proceedings, 1988 IEEE International Conference on Robotics and Automation*, pages 1290–1295. IEEE Press, 1989.

Atlas, L., Cohn, D., Ladner, R., El-Sharkawi, M.A., Marks II, R.J., Aggoune, M.E. & Park, D.C. (1990). Training connectionist networks with queries and selective sampling. In *Advances in Neural Information Processing Systems 2*, pages 567–573. Morgan Kaufmann.

Bard, C., Laugier, C., Milesi, C. & Troccaz, J. (in press). Achieving dextrous grasping by integrating planning and vision based sensing. *The International Journal of Robotics Research*, 1995.

Bard, C., Troccaz, J. & Vercelli, G. (1991). Shape analysis and hand preshaping for grasping. In *IEEE/RSJ International Conference on Intelligent Robots and Systems*, pages 64–69.

Bennett, S (1991). Planning to address uncertainty: An incremental approach employing learning through experience. In *Proceedings of the DARPA Workshop on Advanced Planning and Control Strategies*, pages 313–324.

Berger, J.O. (1985). *Statistical decision theory and Bayesian analysis*. Springer-Verlag, NY.

Box, G.E.P. & Draper, N.R. (1969). *Evolutionary Operation: a Statistical Method for Process Improvement*. Wiley.

Breiman, L., Friedman, J.H., Olshen, R.A. & Stone, C.J. (1984). *Classification and regression trees*. Wadsworth International, Belmont, CA.

Cohn, D.A. (1994). Queries and exploration using optimal experiment design. In *Neural Information Processing Systems 6*, Denver, CO. Morgan-Kauffman.

Cutkosky, M.R. (1989). On grasp choice, grasp models and the design of hands for manufacturing tasks. *IEEE Journal of Robotics and Automation*, 5(3):269–279.

Dunn, G.B. & Segen, J. (1988). Automatic discovery of robotic grasp configuration. In *Proceedings, 1988 IEEE International Conference on Robotics and Automation*, pages 396–401.

Gittins, J.C. (1988). *Multi-armed Bandit Allocation Indices*. John Wiley and Sons, Chichester.

Gullapalli, V., Barto, A.G. & Grupen, R.A. (1994). Learning admittance mappings for force-guided assembly. In *Proceedings, 1994 IEEE International Conference on Robotics and Automation*, pages 2633–2638.

Howe, R.D., Popp, N., Akella, P., Kao, I. & Cutkowsky, M.R. (1990). Grasping, manipulation and control with tactile sensing. In *Proceedings, 1990 IEEE International Conference on Robotics and Automation*, pages 1258–1263.

Iberall, T. Jackson, J., Labbe, L. & Zampano, R.(1988). Knowledge-based prehension: Capturing human dexterity. In *Proceedings, 1988 IEEE International Conference on Robotics and Automation*.

Jeannerod, M. & Biguer, B. (1981). *Advances in the analysis of visual behavior*, chapter Visuomotor mechanisms in reaching within extrapersonal space. MIT Press, Boston.

Jeannerod, M. (1988). *The neural and behavioural organization of goal-directed movements*. Clarendon Press, Oxford Science Publications, Oxford.

Kaelbling, L.P. (1990). *Learning in Embedded Systems*. PhD thesis, Stanford University, Stanford, CA. Dept. of Computer Science.

Kamon, I., Flash, T. & Edelman, S. (1994). Learning to grasp using visual information. Technical Report, The Weizmann Institute of Science, Rehovot, Israel.

Kang, S. & Ikeuchi, K. (1993). Toward automatic robot instruction from perception-recognizing a grasp from observation. *IEEE Journal of Robotics and Automation*, 9(4):432–443.

Kuniyoshi, T., Inaba, M. & Inoue, H. (1989). Teaching by showing: Generating robot programs by visual observation of human performance. In *Proceedings of 20th Internation Syposium on industrial robotics*, pages 119–126.

Larson, R. & Marx, M. (1986). *An introduction to mathematical statistics*. Prentice Hall, Englewood Cliffs, NJ.

Liu, H., Iberall, T. & Bekey, G.A. (1989). The multi-dimensional quality of tasks requirements of task requirements for dextrous hand control. In *IEEE International Conference on Robotics and Automation*, pages 452–457.

Moore, A.W. (1990). Acquisition of dynamic control knowledge for a robotic manipulator. In *Proceedings of the seventh international conference on machine learning*.

Moore, A.W. (1991a). Fast, robust adaptive control by learning only forward models. In *Advances in Neural Information Processing Systems*. Morgan Kauffman.

Moore, A.W. (1991b). Variable resolution dynamic programming: Efficiently learning action maps in multivariate real-valued state-spaces. In *8th International Workshop on Machine Learning*. Morgan Kauffman.

Moreno, C.W. & Yunker, S.B. (1992). Ultramax: Continuous process improvement through sequential optimization. Technical report, Electric Power Research Institute.

Murthy, S., Kasif, S., Salzberg, S. & Beigel, R. (1993). Oc1: Randomized induction of oblique decision trees. In *Proceedings of the 11th National Conference on Artificial Intelligence*, pages 322–327, Menlo Park, CA, USA. AAAI.

Napier, J.R. (1956). The prehensile movements of the human hand. *Journal of Bone and Joint Surgery*, 38b(4):902–904.

Pook, P.K. & Ballard, D.H. (1993). Recognizing teleoperated manipulations. In *Proceedings, 1993 IEEE International Conference on Robotics and Automation*, pages 578–585.

Quinlan, J.R. (1986). *Learning efficient classification procedures and their application to chess end-games*, pages 463–481. Morgan-Kauffman, Los Altos, CA.

Quinlan, R. (1992). *C4.5: Programs for Machine Learning*. Morgan-Kaufmann, Los Altos, CA.

Rumelhart, D.E., Hinton, G.E. & Williams, R. (1986). Learning internal representations by back-propagating errors. *Nature*, 323:533–536.

Salganicoff, M. (1992). *Learning and Forgetting for Perception-Action: A Projection-Pursuit and Density-Adaptive Approach*. PhD thesis, University of Pennsylvania. Dept. of Computer and Information Science.

Salganicoff, M. (1993). Density-adaptive learning and forgetting. In *Proceedings of the Tenth International Conference on Machine Learning*. Morgan-Kauffman.

Salganicoff, M. & Bajcsy, R. (1992). Robot sensorimotor learning in continuous domains. In *1992 IEEE Conference on Robotics and Automation*, Nice, France.

Salganicoff, M. & Ungar, L.H. (1995). Active exploration and learning in real-valued spaces using multi-armed bandit allocation indices. In *Twelfth International Machine Learning Conference*.

Schlesinger, G. (1916). *Der Mechanische Aufbau der Kunstlichen Glieder*. Springer, Berlin.

Schneider, J. (1993). Open loop motor skill learning. In *AAAI Fall Symposium Series: Machine Learning in Vision: What Why and How?*, Raleigh, N.C.

Solina, F. & Bajcsy, R. (1990). Recovery of parametric models from range images: The case for superquadrics with global deformations. *IEEE Transactions on Pattern Analysis and Machine Intelligence*, 12(2):131–147.

Stansfield, S. (1990). Knowledge-based robotic grasping. In *Proceedings, 1990 IEEE International Conference on Robotics and Automation*, pages 1270–1275.

Stansfield, S.A. (1992). Connectionist and neural net implementations of a robotic grasp generator. In *Proceedings of SPIE - Engineering Applications of Artificial Intelligence X: Machine Vision and Robotics*, volume 1708, pages 375–388, Bellingham, WA, USA. Int. Soc. for Optical Engineering.

Sutton, R.S. (1988). Learning to predict by the methods of temporal differencing. *Machine Learning*, 3:9–44.

Sutton, R.S. (1990). Integrated architectures for learning planning and reacting based on approximating dynamic programming. In *Proceedings of the Seventh International Conference on Machine Learning*. Morgan-Kauffman.

Tan, M. (1990). A cost-sensitive learning system for sensing and grasping objects. In *Proceedings, 1990 IEEE International Conference on Robotics and Automation*. IEEE Press.

Tan, M. (1993). Cost-sensitive learning of classification knowledge and its applications in robotics. *Machine Learning*, 13(1):7–33.

Taylor, M., Blake, A. & Cox, A. (1994). Visually guided grasping in 3D. In *Proceedings, 1994 IEEE International Conference on Robotics and Automation*, pages 761–766.

Thrun, S. & Moller, K. (1992). Active exploration in dynamic environments. In *Neural Information Processing Systems 4*, pages 531–538. Morgan-Kauffman.

Thrun, S.B. (1992). The role of exploration in learning control with neural networks. In *Handbook of Intelligent Control: Neural, Fuzzy and Adaptive Approaches*. Van Nostrand Rheinhold, Florence KY.

Tsikos, C. & Bajcsy, R. (1991). Segmentation via manipulation. *IEEE Transactions on Robotics and Automation*, 7(3):306–319.

Utgoff, P. (1988). ID-5: An incremental ID-3. In *Proceedings of the Fifth International WOrkshop on Machine Learning*. Morgan-Kauffmam.

Watkins, C. J. C. H. (1989). *Learning from delayed rewards*. PhD thesis, King's College, Cambridge.

Machine Learning, 23, 279–303 (1996)

Purposive Behavior Acquisition for a Real Robot by Vision-Based Reinforcement Learning

MINORU ASADA asada@robotics.ccm.eng.osaka-u.ac.jp
SHOICHI NODA
SUKOYA TAWARATSUMIDA
KOH HOSODA
Dept. of Mech. Eng. for Computer-Controlled Machinery
Osaka University, 2-1, Yamadaoka, Suita, Osaka 565, Japan

Editors: Judy A. Franklin, Tom M. Mitchell, and Sebastian Thrun

Abstract. This paper presents a method of vision-based reinforcement learning by which a robot learns to shoot a ball into a goal. We discuss several issues in applying the reinforcement learning method to a real robot with vision sensor by which the robot can obtain information about the changes in an environment. First, we construct a state space in terms of size, position, and orientation of a ball and a goal in an image, and an action space is designed in terms of the action commands to be sent to the left and right motors of a mobile robot. This causes a "state-action deviation" problem in constructing the state and action spaces that reflect the outputs from physical sensors and actuators, respectively. To deal with this issue, an action set is constructed in a way that one action consists of a series of the same action primitive which is successively executed until the current state changes. Next, to speed up the learning time, a mechanism of *Learning from Easy Missions* (or LEM) is implemented. LEM reduces the learning time from exponential to almost linear order in the size of the state space. The results of computer simulations and real robot experiments are given.

Keywords: reinforcement learning, vision, learning from easy mission, state-action deviation

1. Introduction

Realization of autonomous agents that organize their own internal structure in order to take actions towards achieving their goals is the ultimate goal of AI and Robotics. That is, the autonomous agents have to learn. Reinforcement learning has recently been receiving increased attention as a method for robot learning with little or no *a priori* knowledge and higher capability of reactive and adaptive behaviors (Connel & Mahadevan, 1993). In the reinforcement learning method, a robot and its environment are modeled by two synchronized finite state automatons interacting in discrete time cyclical processes. The robot senses the current state of the environment and selects an action. Based on the state and the action, the environment makes a transition to a new state and generates a reward that is passed back to the robot. Through these interactions, the robot learns a purposive behavior to achieve a given goal.

Although the role of reinforcement learning is very important to realize autonomous systems, the prominence of that role is largely dependent on the extent to which the learning can be scaled to solve larger and more complex robot learning tasks. Many researchers in the field of machine learning have been concerned with the convergence time of the learning, and have developed methods to speed it up. They have also extended these techniques from solving single goal tasks to multiple goal ones (Sutton, 1992). However, almost all of them have only shown computer simulations in which they assume ideal sensors and actuators, where they can easily construct the state and action spaces consistent with each other. A typical example is the 2-D grid environment in which the robot can take an action of going forward, backward, left, or right, and its state is encoded by the coordinate of the grid (i.e., an absolute (global) positioning system is assumed). Although the uncertainties of sensor and actuator outputs are considered by a stochastic transition model in the state space, such a model cannot account for the accumulation of sensor errors in estimating the robot position. Further, from the viewpoint of real robot applications, we should construct the state space so that it can reflect the outputs of the physical sensors which are currently available and can be mounted on the robot.

Some applications are recently reported to control robot arms (Saito & Fukuda, 1994) or mobile robots (Fagg, Lotspeich, & Bekey, 1994) in which the initial controller and the correct reward function are given in advance. Therefore, the robot learns the control policy given a great deal of knowledge about the environment and itself. We intend to apply the reinforcement learning algorithm to the task of purposive behavior acquisition in the real world with less knowledge about the environment and the robot.

Mahadevan and Connel (1993) proposed a method of rapid task learning on a real robot. They separated a pushing task into three subtasks of "finding a box", "pushing a box", and "getting unwedged", and applied Q-learning, a widely used reinforcement learning method, to each of them. Since only proximity sensors such as bumper and sonar sensors are used, the acquired behaviors are limited to local ones and therefore these behaviors are not suitable for more global and goal-directed tasks such as carrying a box to a specified location. For such tasks, visual sensors could be more useful because they might be able to capture the image of the goal in a distant place. However, there are very few examples of use of visual information in reinforcement learning [1], probably because of the cost of visual processing.

In this paper, we present a method of vision-based reinforcement learning by which a robot learns to shoot a ball into a goal. The robot does not need to know any parameters of the 3-D environment or its kinematics/dynamics. The image captured from a single TV camera mounted on the robot is the only source of information on the changes in an environment. Image positions and sizes of the ball and the goal are used as a state vector. We discuss several issues from a viewpoint of robot learning: a) coping with a "state-action deviation" problem which occurs in constructing the state and action spaces in accordance with outputs from the physical sensors and actuators, and b) starting with easy missions (rather than task decomposition) for rapid task learning.

The remainder of this article is structured as follows: In the next section, we explain the task and assumptions, and give a brief overview of Q-learning. Next, we show how to construct the state and action spaces for the task at hand, and describe ways to reduce the learning time by Learning from Easy Missions (or LEM) mechanism. Finally, we show the experimental results obtained by computer simulations and using the real robot system, followed by a discussion and concluding remarks.

2. Task and Assumptions

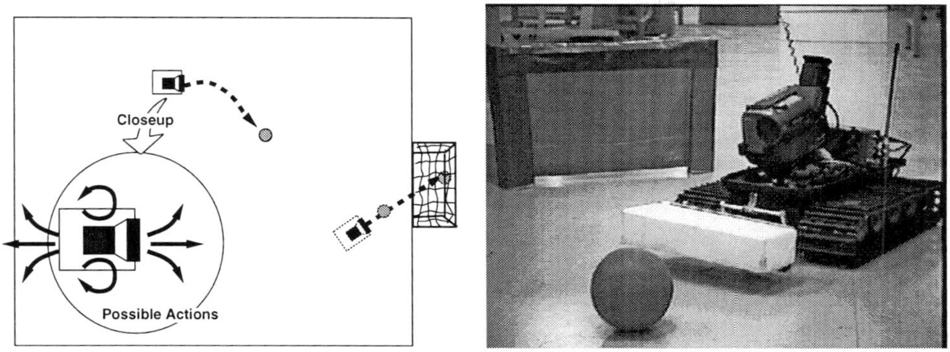

(a) The task is to shoot a ball into a goal. (b) A picture of the radio-controlled vehichle.

Figure 1. Task and our real robot.

The task for a mobile robot is to shoot a ball into a goal as shown in Figure 1(a). The problem we address here is how to develop a method which automatically acquires strategies for doing this. We assume that the environment consists of a ball and a goal; the mobile robot has a single TV camera; and that the robot does not know the location/size of the goal, the size/weight of the ball, any camera parameters such as the focal length and tilt angle, or the kinematics/dynamics of itself. Figure 1(b) shows a picture of the real robot with a TV camera (Sony handy-cam TR-3) used in the experiments.

3. Q-learning

Before getting into the details of our system, we will briefly review the basics of Q-learning. For a more thorough treatment, see (Watkins, 1989). We follow the explanation of Q-learning by Kaelbling (Kaelbling, 1993).

We assume that the robot can discriminate the set $\boldsymbol{S}$ of distinct world states, and can take the set $\boldsymbol{A}$ of actions on the world. The world is modeled as a Markov process, making stochastic transitions based on its current state and the action taken by the robot. Let $T(s, a, s')$ be the probability of transition to the state s' from the current state-action pair (s, a). For each state-action pair (s, a), the *reward* $r(s, a)$ is defined.

The general reinforcement learning problem is typically stated as finding a policy [2] that maximizes the discounted sum of rewards received over time. This sum is called the $return$ and is defined as:

$$\sum_{n=0}^{\infty} \gamma^n r_{t+n}, \tag{1}$$

where r_t is the reward received at step t given that the agent started in state s and executed policy f. γ is the discounting factor, it controls to what degree rewards in the distant future affect the total value of a policy. The value of γ is usually slightly less than 1.

Given definitions of the transition probabilities and the reward distribution, we can solve for the optimal policy, using methods from dynamic programming (Bellman, 1957). A more interesting case occurs when we wish to simultaneously learn the dynamics of the world and construct the policy. Watkin's Q-learning algorithm gives us an elegant method for doing this.

Let $Q^*(s,a)$ be the expected return or *action-value function* for taking action a in a situation s and continuing thereafter with the optimal policy. It can be recursively defined as:

$$Q^*(s,a) = r(s,a) + \gamma \sum_{s' \in S} T(s,a,s') \max_{a' \in A} Q^*(s',a'). \tag{2}$$

Because we do not know T and r initially, we construct incremental estimates of the Q-values on-line. Starting with $Q(s,a)$ equal to an arbitrary value (usually 0), every time an action is taken, the Q-value is updated as follows:

$$Q(s,a) \Leftarrow (1-\alpha)Q(s,a) + \alpha(r(s,a) + \gamma \max_{a' \in A} Q(s',a')). \tag{3}$$

where r is the actual reward value received for taking action a in a situation s, s' is the next state, and α is a learning rate (between 0 and 1).

4. Construction of State and Action Sets

Traditional notions of state in the existing applications of the reinforcement learning algorithms fit nicely into deterministic state transition models (e.g. one action is forward, backward, left, or right, and the states are encoded by the locations of the agent). However, this is not always the case in the real world, where everything changes asynchronously (Mataric, 1994). Thus, we need to have the following principles for the construction of state and action spaces.

- Natural segmentation of the state and action spaces: The state (action) space should reflect the corresponding physical space in which a state (an action) can be perceived (taken).

- Real-time vision system: Physical phenomena happen continuously in the real world. Therefore, the sensor system should monitor the changes of the environment in real time. This means that the visual information should be processed in video frame rate (33ms).

The state and action spaces are not discrete but continuous in the real world, therefore it is difficult to construct the state and action spaces in which one action always corresponds to one state transition. We call this the **"state-action deviation problem"** as one of the so-called "perceptual aliasing problem" (Whitehead & Ballard, 1990) (i.e., a problem caused by multiple projections of different actual situations into one observed state). The perceptual aliasing problem makes it very difficult for a robot to take an optimal action. In this section, we first show how to construct the state and action spaces, and then how to cope with the state-action deviation problem.

4.1. Construction of Each Space

(a) a state set $\boldsymbol{S}$

The image, supposed to capture the ball and/or the goal, is the only source of information the robot can obtain about the environment. The ball image is classified into 9 sub-states, combinations of three classifications of positions (left, center, or right) and three types of sizes (large (near), middle, or small (far)). The goal image has 27 sub-states, combinations of three properties each of which is classified into three categories (see Figure 2). Each sub-state corresponds to one posture of the robot towards the goal, that is, position and orientation of the robot in the field. In addition to these 243 (27 $\times$ 9) states, we add other states such as the cases in which only the ball or only the goal is captured in the image. In all, we have 319 states in the set $\boldsymbol{S}$.

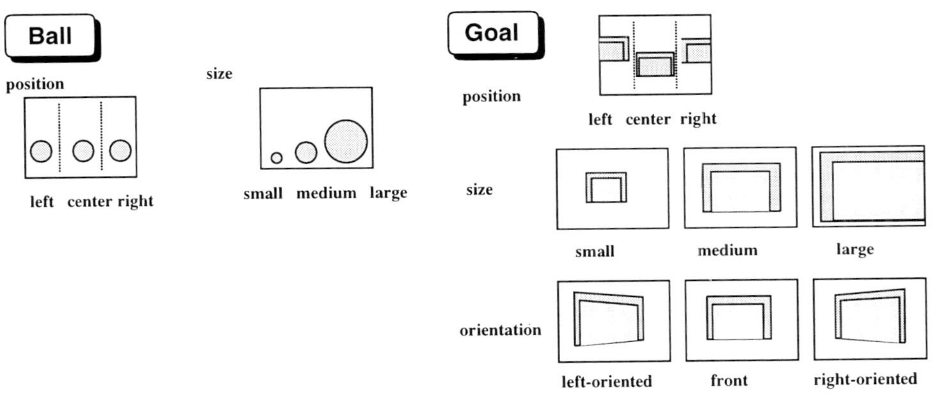

Figure 2. The ball sub-states and the goal sub-states

After some simulations, we realized that as long as the robot captures the ball and the goal positions in the image it succeeds in shooting a ball into the goal. However, once it cannot view the ball, the robot moves randomly because it does not know in which direction it should move to find the ball. This occurs because "ball-lost" is just one state, therefore the robot cannot discriminate between the various directions in which the ball may be lost. Thus, we separate the ball-lost state into two states; the ball-lost-into-right and ball-lost-into-left states. Similarly, we set up the goal-lost-into-right and goal-lost-into-left states. These additional classifications improved the robot behavior.

(b) an action set $\boldsymbol{A}$

The robot can select an action to be taken in the current state of the environment. The robot moves around using a PWS (Power Wheeled Steering) system with two independent motors. Since we can send the motor control command to each of the two motors separately, we construct the action set in terms of two motor commands ω_l and ω_r, each of which has 3 sub-actions, forward, stop, and back. All together, we have 9 actions in the action set $\boldsymbol{A}$.

(c) a reward and a discounting factor γ

We assign the reward value to be 1 when the ball is kicked into the goal and 0 otherwise. This makes the learning very time-consuming. Although adopting a reward function in terms of distance to the goal state makes the learning time much shorter in this case, it seems difficult to avoid the local maxima of the action-value function Q.

A discounting factor γ is used to control to what degree rewards in the distant future affect the total value of a policy. In our case, we set the value at slightly less than 1 ($\gamma = 0.8$).

4.2. Solving A State-Action Deviation Problem

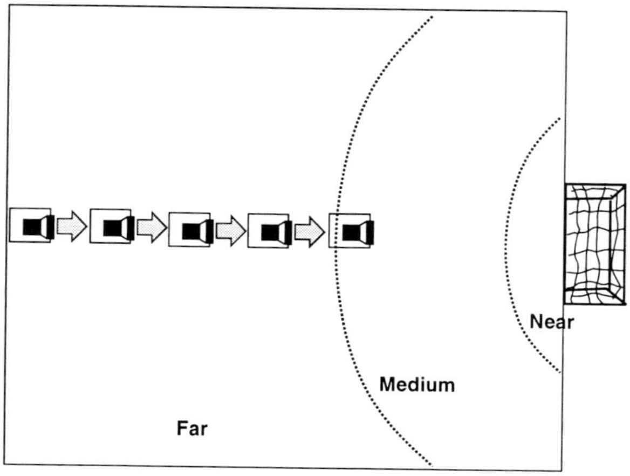

Figure 3. A state-action deviation problem

In the previous section, we constructed the state space so that the position and the size of the ball or goal are naturally and coarsely classified into each state. Due to the peculiarity of visual information, that is, a small change near the observer results in a large change in the image and a large change far from the observer may result in a small change in the image, one action does not always correspond to one state transition. We call this the **"state-action deviation problem"**: Figure 3 indicates this problem, the area representing the state "the goal is far" is large, therefore the robot frequently returns to this state if the action is forward. This is highly undesirable because the variations in the state transitions is very large, consequently the learning does not converge correctly.

To avoid this problem, we reconstruct the action space as follows. Each action defined in **4.1** is regarded as an action primitive. The robot continues to take one action primitive at a time until the current state changes. This sequence of the action primitives is called an action. In the above case, the robot takes a forward motion many times until the state "the goal is far" changes into the state "the goal is medium". The number of action primitives needed for one state change is not used in the algorithm. Once the state has changed, we update the action-value function by Eqn.(3).

5. Learning from Easy Missions

In order to improve the learning rate, the whole task was separated into different parts in (Connel & Mahadevan, 1993). By contrast, we do not decompose the whole task into subtasks of finding, driblling, and shooting a ball. Instead, we first used a monolithic approach. That is, we place the ball and the robot at arbitrary positions. In almost all the cases, the robot crossed over the field line without shooting the ball into the goal. This means that the learning did not converge after many trials (execution time of one week on SGI Elan with R4000). This situation resembles a case where a small child tries to shoot a ball into a goal, but (s)he cannot imagine in which direction and how far the goal is because a reward is received only after the ball is kicked into the goal. Further, (s)he does not know which action is to be selected. This is the famous *delayed reinforcement* problem due to no explicit teacher signal that indicates the correct output at each time step. To avoid this difficulty, we construct the learning schedule such that the robot can learn in easy situations at the early stages and later on learn in more difficult situations. We call this *Learning from Easy Missions* (or LEM).

In the following, we first show the complexity analysis of LEM assuming that the agent knows the order of state transitions completely. Although it does not make any sense if we know it because the learning is no longer necessary, we intend to show the difference in learning times between with and without LEM. Then, we describe how to cope with two problems in implementing the LEM in real tasks: how to decide which state is easier to achieve the goal and when to shift from easy situations to more difficult ones.

5.1. Complexity analysis of LEM

Figure 4. The simplest state space.

We roughly estimate the time complexity for LEM following the complexity analysis by Whitehead (Whitehead, 1991). In order to show the difference in the learning time with and without LEM, we assume complete knowledge of the order of the states to the goal. However, as mentioned above, the LEM does not always need completeness of the knowledge.

We assume a "homogeneous" state space uniformly k-bounded with polynomial width of depth k, and zero-initialized Q-learning with a problem solving task. The problem solving task is defined as any learning task where the system receives a reward only upon entering a goal state. Further, we assume that state transition is deterministic and the robot can take one of m kinds of actions with equal probability. In order to figure out how many steps are needed for Q-learning to converge, we use $O(k)$ state space and simplify the convergence such that the action-value function in each state converges if it is updated from the initial value (0).

Figure 4 shows an example of such a state space. Since we consider a problem solving task, the unbiased Q-learning takes a long time. From the above formulation, it needs m trials [3] to transit from the initial state S_k to the state S_{k-1} in the worst case, therefore it takes m^k trials to achieve the goal for the first time in the worst case, and the action-value function for only the state S_1 is updated. Next, it needs m^{k-1} trials to update the action-value function for the state S_2, and in total it needs $(m^k + m^{k-1} + \cdots + m)$ trials for the action value function to converge for all the states. Therefore, the unbiased Q-learning can be expected to take execution time exponential in the size of k (Whitehead, 1991).

For the Learning from Easy Missions algorithm, we first place the agent at the state S_1 and then try to make it achieve the goal. In the worst case, it takes m trials. Then, we place the agent at the state S_2 and repeat the process. In the worst case, it needs $m \times k$ trials for the action-value function to converge. Therefore, the LEM algorithm requires time linear in the size of k.

5.2. Implementing LEM

In actual situations like our task, the state transition is stochastic, and the state space is not homogeneous. Therefore, it seems difficult to correctly decide which state is an easy one to achieve the goal and when to shift the initial situations into more difficult ones. In the following, we show how to cope with this problem.

(a) Rough ordering of easy situations

Generally, it seems difficult to know which state is easier than others to achieve the goal, unless *a priori* knowledge on ease of achieving the goal is given. However, in real applications, we can assume that the sensor space is smooth and the goal state is known. Therefore, the agent roughly knows the order of state transitions such as the size of a goal or a ball in the image regardless of other sub-states in our task:"small" → "medium" → "large" → "goal state". Similarly, the position of the goal and ball are also ordered such as "left (or right)" → "center" → "goal state". However, the orientation of the goal is not ordered because the agent can achieve the goal from any direction.

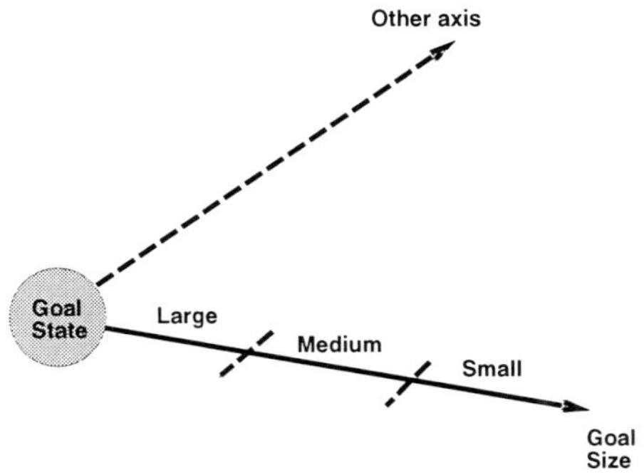

Figure 5. State space part of which divided into sub-states

This ordering is possible if the state space is categorized into sub-states such as ball size, ball position, and so on. Figure 5 shows a state space in which a number of states are ordered along the sub-state axis. The LEM paradigm can be applied when such a sub-state axis exists or partial ordering along the axis is known. Let n and m be the size of the state space (the number of states) and the number of ordered sets (e.g., 3, when the sub-state axis is the goal size and states are "small", "medium", and "larg e"). When the agent learns without LEM it takes $O(e^n)$. If we apply the LEM with m ordered state sets, it would take $O(m \times \sqrt[m]{e^n})$.

(b) When to shift?

Another problem in implementing the LEM paradigm is to decide when to shift the initial situations into more difficult ones. From the above discussion, we suppose that the agent knows the ordering of situations along one sub-state axis according to the closeness of the goal state. We call a set of states $\mathbf{S_1}$ that is closest to the goal along the current sub-state axis regardless of other sub-state axes, the set of second closest states $\mathbf{S_2}$, the set of third closest states $\mathbf{S_3}$, and so on. Shifting to the next set occurs when

$$\Delta Q_t(\mathbf{S_k}, a) < \epsilon, \tag{4}$$

where

$$\Delta Q_t(\mathbf{S_k}, a) = \sum_{s \in \mathbf{S_k}} \left| \max_{a \in \mathbf{A}} Q_t(s, a) - \max_{a \in \mathbf{A}} Q_{t-\Delta t}(s, a) \right| \qquad (k = 1, 2, 3, \cdots). \tag{5}$$

Δt indicates a time interval for a number of steps of state changes.

The closer to zero ϵ is in inequality (4), the earlier Q converges to Q^* in a deterministic world, because it takes longer time to converge Q-values in the state set $\mathbf{S_k}$ if we shift the initial states to the state set $\mathbf{S_{k-1}}$ earlier than the case where $\epsilon \approx 0$. It is, however, not always true in a non-deterministic world that $\Delta Q_t(\mathbf{S_k}, a) = 0$ when the learning is at its final stage. Therefore, we have to set up an appropriate value for ϵ.

We suppose that the current state set $\mathbf{S_{k-1}}$ cannot transit to any state but is limited to transitions to states in the neighborhood of sets $\mathbf{S_{k-2}}$ and $\mathbf{S_k}$. When the learning has converged sufficiently, we have the following from Eqn. (3).

$$\Delta Q(s,a) = \left| \alpha(r(s,a) + \gamma \max_{a' \in \mathbf{A}} Q(s', a') - Q(s,a)) \right| . \tag{6}$$

Since $r(s,a) = 0$ except the goal state,

$$\Delta Q(s,a) = \begin{cases} \alpha\frac{1-\gamma}{\gamma}Q(s,a), & \text{for } s' \in \mathbf{S_{k-2}} \\ 0, & \text{for } s' \in \mathbf{S_{k-1}} \\ \alpha(1-\gamma)Q(s,a). & \text{for } s' \in \mathbf{S_k} \end{cases} \tag{7}$$

If $Q(s,a)$ has converged, then:

$$\Delta Q(s,a) \leq \alpha\frac{1-\gamma}{\gamma}Q(s,a).$$

Taking summation over elements in the state set $\mathbf{S_k}$ leads to:

$$\Delta Q(\mathbf{S_k}, a) \leq \alpha\frac{1-\gamma}{\gamma}\sum_{s \in \mathbf{S_k}} \max_{a \in \mathbf{A}} Q(s,a).$$

Note that the sampling time Δt should be long enough to check the convergence exactly without being affected by small fluctuations of the summation of Q-values, but also short enough to reduce the total learning time. Since the above inequality supposes that the Q-values for all $s \in \mathbf{S_k}$ are updated, an adequate Δt should be longer than $|\mathbf{S_k}|$ attempts [4] to make attempts starting from every $s \in \mathbf{S_k}$. Following the above discussions, we adopt:

$$\epsilon = \alpha\frac{1-\gamma}{\gamma}\sum_{s \in \mathbf{S_k}} \max_{a \in \mathbf{A}} Q(s,a). \tag{8}$$

From the above, the LEM algorithm is formulated as follows:

1. Find a sub-state axis along which rough ordering of state transition to a goal state is known.

2. Let $\mathbf{S_i}$ $(i = 1, 2, .., m)$ be a set of states that is closest to the goal along the current sub-state axis regardless of other sub-state axes, a set of second closest states, a set of third closest states, and so on.

3. Set $i = 1$.

4. Repeat many attempts starting with various kinds of initial situations which are classified into the state set $\mathbf{S_i}$ until the inequality (4) is satisfied (Action selection is random).

5. If $i = m$, then terminate. Else, $i = i + 1$ and go to step 4.

5.3. Simulation results

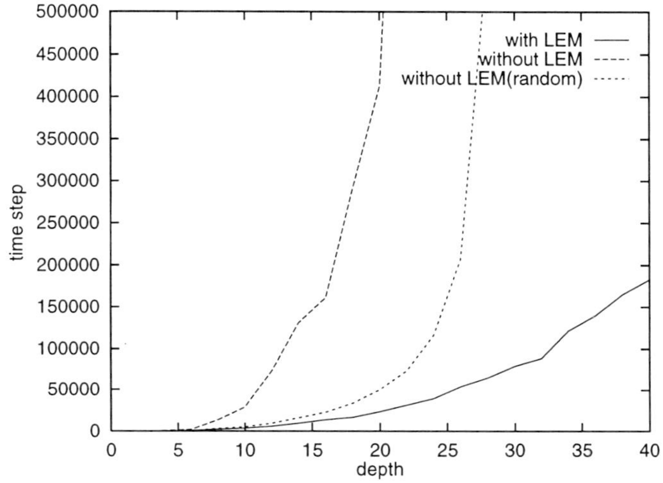

Figure 6. Search time complexity in terms of the size of state space.

To start with a simple example, we consider the grid world problem. In this example, the robot is free to roam about a bounded 2-dimensional grid $n \times n$. It can move to one of four principle directions, left, right, up, or down. The robot is reset after it reaches the goal or crosses over the grid world boundary. In order to simulate the stochastic state transitions, the robot can move to a desired grid with probability 3/4 but stays at the current grid with probability 1/4. The goal is set at the top right corner and the state is encoded as a coordinate of the grid. We assign a reward value 1 when the robot achieves the goal and 0 otherwise. The depth of the state space, the maximum distance to the goal of the optimal path, is $2(n-1)$.

We implemented the LEM as follows. The state set $\mathbf{S_1}$ consists of two states (each of which is a grid neighbor to the goal), $\mathbf{S_2}$ consists of three states (each of which is a grid neighbor to $\mathbf{S_1}$), and so on. We shift the initial grids of $\mathbf{S_k}$ to that of $\mathbf{S_{k+1}}$ according to Eqns. (4) and (8) with $\Delta t = |\mathbf{S_j}|$ attempts.

Figure 6 shows a plot of the search time (number of steps) versus maximum distance k, where the one step Q-learning algorithm is applied with the learning rate $\alpha = 0.25$ and the discounting factor $\gamma = 0.9$. As we expected, the search time of the normal Q-learning without LEM (dotted and broken lines) indicates the exponential order in the size of k. The initial position is fixed at the bottom left corner (broken line) or randomly placed (dotted line). In comparison with Q-learning without LEM, the search time with LEM (solid line) is almost linear in the size of k although the curve is a little bit upward because some inadequate attempts wasted much time due to useless search such as backward motions.

6. Experiments

The experiment consists of two parts: first, learning the optimal policy f through the computer simulation, then applying the learned policy to a real situation.

6.1. *Computer simulation*

We performed the computer simulation with the following specifications. The field is a square of 3.0m $\times$ 3.0m. The goal post is located at the center of the top line of the square (see Figure 1) and its height and width are 0.23m and 0.9m, respectively. The robot is 0.31m wide and 0.45m long and kicks a ball of diameter 0.09m. The maximum translation velocity is 1.1m/s, and the maximum angular velocity is 4.8rad/s. The camera is horizontally mounted on the robot (no tilt), and its visual angle is 36 degrees. The velocities of the ball before and after being kicked by the robot is calculated by assuming that the mass of the ball is negligible compared to that of the robot. The speed of the ball is temporally decreased by a factor 0.8 in order to reflect the so-called "viscous friction." The values of these parameters are determined so that they can roughly simulate the real world.

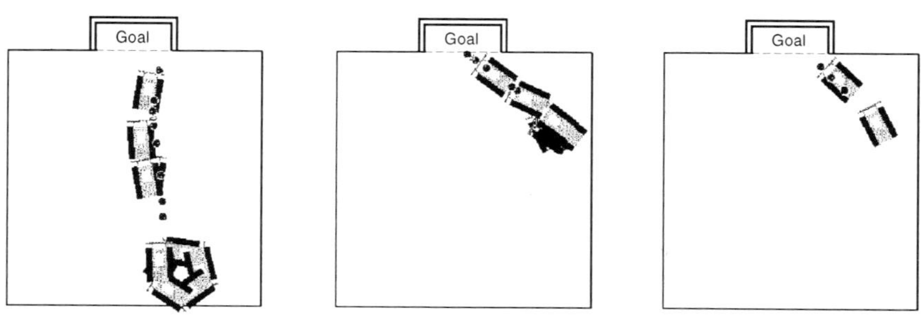

(a) finding, dribbling, and shooting (b) shooting ($\gamma = 0.999$) (c) shooting ($\gamma = 0.6$)

Figure 7. Some kinds of behaviors obtained by our method

Figure 7 shows some kinds of behaviors obtained by our method. In (a), the robot started at a position from where it could not view a ball and a goal, then found the ball by turning, driblled it towards the goal, and finally shot the ball into the goal. This is just a result of learning. We did not decompose the whole task into these three tasks. The difference in the character of robot player due to the discounting factor γ is shown in (b) and (c) in which the robot started from the same position. In the former, the robot takes many steps in order to ensure the success of shooting because of a small discount, while in the latter the robot tries to shoot a ball immediately because of a large discount. In the following experiments, we used the average value of γ 0.8 as an appropriate discount.

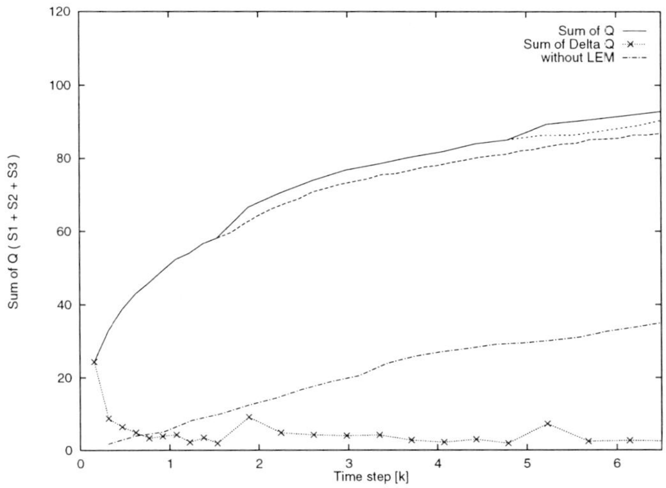

Figure 8. Change of the sum of Q-values with LEM in terms of goal size

We applied the LEM algorithm to the task in which $\mathbf{S_i}$ (**i**=1,2, and 3) correspond to the state sets of "the goal is large", "medium", and "small", respectively, regardless of the orientation and the position of the goal, and the size and position of the ball. ϵ is determined by Eqns. (4) and (8). In order to ensure the convergence, we fix $\Delta t = 3000$ that is much larger than $|\mathbf{S_j}|$.

Figure 8 shows the changes of the summations of Q-values with and without LEM, and ΔQ. The axis of time step is scaled by M (10^6), which corresponds to about 9 hours in the real world since one time step is 33ms. The solid and broken lines indicate the summations of the maximum value of Q in terms of action in states $\in \mathbf{S_1} + \mathbf{S_2} + \mathbf{S_3}$ with and without LEM, respectively. The Q-learning without LEM was implemented by setting initial positions of the robot at completely arbitrary ones. Evidently, the Q-learning with LEM is much better than that without LEM. The broken line with empty squares indicates the change of $\Delta Q(\mathbf{S_1} + \mathbf{S_2} + \mathbf{S_3}, a)$. Two arrows indicate the time steps (around 1.5M and 4.7M) when a set of the initial states changed from $\mathbf{S_1}$ to $\mathbf{S_2}$ and from $\mathbf{S_2}$ to $\mathbf{S_3}$, respectively. Just after these steps, ΔQ drastically increased, which means the Q-values in the inexperienced states are updated. The coarsely and finely dotted lines expanding from the time steps indicated by the two arrows show the curves when the initial positions were not changed from $\mathbf{S_1}$ to $\mathbf{S_2}$, nor from $\mathbf{S_2}$ to $\mathbf{S_3}$, respectively. This simulates the LEM with partial knowledge. If we know only the easy situations ($\mathbf{S_1}$), and nothing more, the learning curve follows the finely dotted line in Figure 8. The summation of Q-values is slightly less than that of the LEM with more knowledge, but much better than that without LEM.

Similarly, we applied the LEM algorithm with other orderings in terms of ball position. Figure 9 show a graph similar to Figure 8. Since the ball position is classified into two categories such as "center" and "left (or right)", the number of arrows indicating the time step to shift from a set of easier states to a set of more difficult ones is just one. Although the difference of LEM with partial and complete knowledge on the ordering along this sub-axis is smaller than the case of the goal size, the usefulness of the LEM algorithm can be seen.

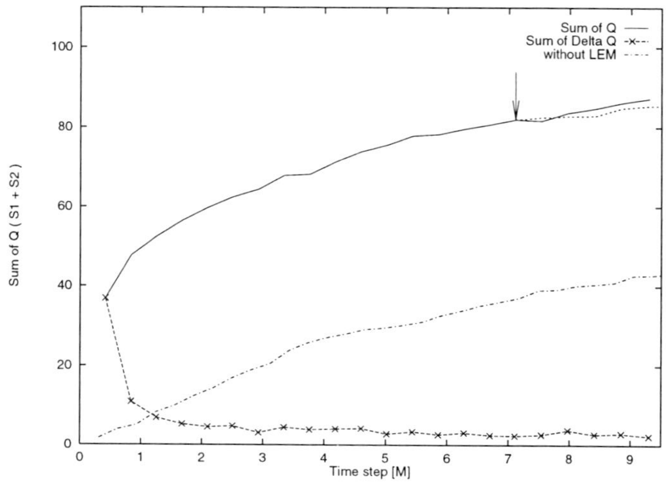

Figure 9. Change of the sum of Q-values with LEM in terms of ball position

6.2. Experimental results on a real robot

Figure 10 shows a configuration of the real mobile robot system. The image taken by a TV camera mounted on the robot is transmitted to a UHF receiver and processed by Datacube MaxVideo 200, a real-time pipeline video image processor. We constructed the radio control system of the vehicle (Inaba, 1993). The image processing and the vehicle control system are operated by VxWorks OS on MC68040 CPUs which are connected with host Sun workstations via Ether net. We have shown a picture of the real robot in Figure 1(b).

In order to simplify and speed up the image processing time, we painted the ball in red and the goal in blue. The input NTSC color video signal is first converted into HSV color components in order to make the extraction of the ball and the goal easy. Then, the image size is reduced to further speed up the image processing time, and the boundaries of the ball and goal regions are extracted for the state discrimination. The results from the image processing operations are sent to the host CPU which decides an optimal action for the current state. Figure 11 shows an example of the input image captured by the robot and the result of the ball and the goal detection for the input image, respectively. Their positions, sizes, and the goal orientation are calculated in real time (in total, 33ms). The image processing program can handle the occlusion by using knowledge of the object shapes such as a sphere and a rectangle.

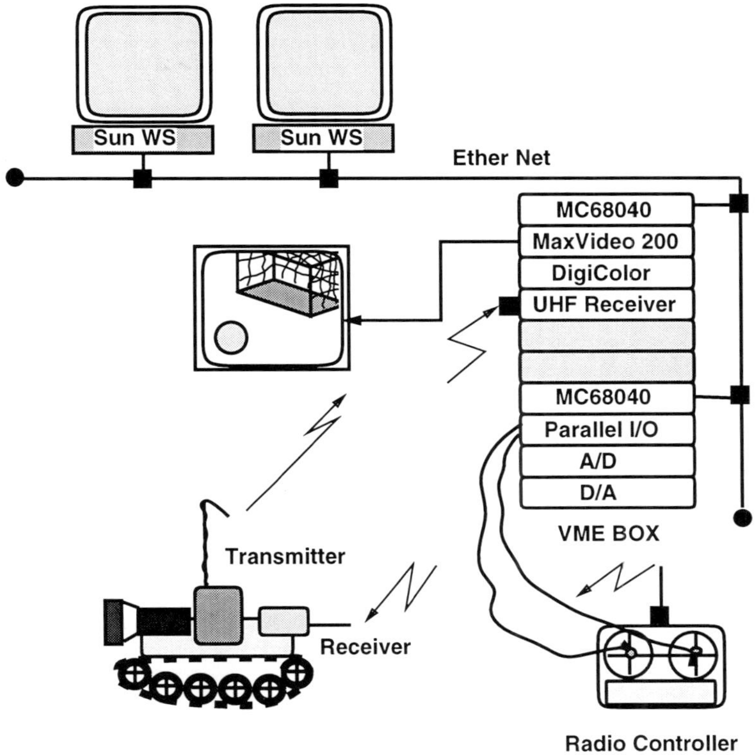

Figure 10. A configuration of the real system.

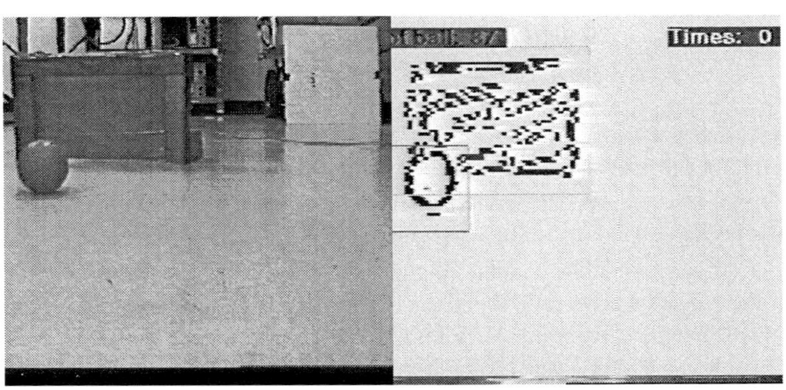

Figure 11. Result of image processing.

We tested the learned policy on the real robot almost 100 times. The shooting rate in the real robot system was about 60% which was about 20% worse than the simulation. The

Figure 12. The robot succeeded in shooting a ball into the goal(I).

main reason for this is that the ball often moves towards unpredictable directions due to the eccentricity of its centroid. The second cause for poorer performance is noise in the image processing results, explained in the following paragraph.

Figure 12 shows a sequence of the images (raster order from top left to bottom right) in which the robot succeeded in shooting a ball into the goal by using our method. The images are taken every five frames (166ms). Table 1 shows the image processing and state mapping result in each time step (33ms) for the sequence of images captured by the robot, examples of which are shown in Figure 12. Each column indicates the time step, the state transition step, the mapped state, the action command, and the number of errors in the state discrimination process, respectively. The mapped state consists of five sub-states: two for the ball position (**L**eft, **C**enter, or **R**ight) and size (large (**N**ear), **M**iddle, or small (**F**ar)), and three for the goal position, size, and orientation (**L**eft-**o**riented, **F**ront-**o**riented, or **R**ight-**o**rinented). "**D**" means a lost state (disappear). Incorrectly mapped sub-states are

Figure 13. The robot failed to shoot a ball into the goal.

marked with "*"s, and the number of these sub-states are shown in the error box. An action command consists of a combination of two independent motor commands (**F**orward, **S**top, or **B**ackward).

Amazingly, the ratio of completely correct mappings is about 60%. Most of the incorrect mappings occur when the size of the ball is misjudged as smaller due to failures in edge detection. As long as the ball and the goal are captured at the center of the image, this does not cause a serious problem because the optimal action is forward in any case. However, the robot fails to shoot a ball when the ball is captured at the right or left of the image, because it misjudges the distance to the ball. Due to the noise of the transmitter, completely incorrect mappings occur at the ratio of 15%. Unless this situation continues two or more

Figure 14. The robot succeeded in shooting a ball into the goal(II).

time steps, this does not cause a serious problem because the robot can take the correct action once the proper state mapping is obtained.

Other examples are shown in Figures 13 and 14. In the former, the robot failed to shoot a ball. The images are taken every 12 frames (about 400ms). First, the robot tried to shoot a ball into a goal, but the ball moved in an unpredicted direction. Then, the robot moved backward, and tried again. Still it could not shoot the ball into the goal, but it seems that the robot had an intention to do that even though it failed. Note that, the shooting is not a programmed action but just a result of learning. In Figure 14, the images are taken every second. First, the robot lost the ball due to noise, and then it turned around to find the ball, and finally it succeeded in shooting.

Table 1. State-Action data

time step	state step	state ball	state goal	action L	action R	# of erros
1	1	(C,F)	(C,F,Fo)	F	F	
2	2	(R*,F)	(C,F,Fo)	F	F	1
3	3	(D*,D*)	(C,F,Ro*)	B	B	3
4	4	(C,F)	(C,F,Lo*)	B	S	1
5	5	(C,F)	(C,F,Fo)	F	F	
6		(C,F)	(C,F,Fo)	F	F	
7		(C,F)	(C,F,Fo)	F	F	
8		(C,F)	(C,F,Fo)	F	F	
9	6	(C,F)	(C,F,Ro*)	B	S	1
10	7	(C,F)	(C,F,Fo)	F	F	
11	8	(C,F)	(R,M,Fo)	F	F	
12	9	(R,F)	(R,M,Fo)	F	F	
13	10	(R,M*)	(R,F*,Lo*)	F	B	3
14	11	(L*,F)	(R,M,Ro*)	F	S	2
15	12	(L*,F)	(R,M,Fo)	F	S	1
16	13	(R,M)	(R,M,Fo)	S	B	
17	14	(C,M)	(C,M,Fo)	F	F	
18	15	(L,M)	(L,M,Fo)	S	F	
19	16	(L,N)	(L,M,Fo)	B	S	
20		(L,N)	(L,M,Fo)	B	S	
21	17	(L,M*)	(L,M,Fo)	S	F	1
22	18	(L,N)	(L,M,Fo)	B	S	
23		(L,N)	(L,M,Fo)	B	S	
24	19	(C,N)	(C,M,Fo)	F	B	
25	20	(C,M)	(C,M,Fo)	F	F	
26		(C,M)	(C,M,Fo)	F	F	
27	21	(C,M)	(C,N,Fo)	F	S	
28	22	(C,M)	(C,M*,Lo*)	F	S	2
29	23	(C,M)	(C,M*,Ro*)	S	B	2
30	24	(C,F)	(D,D,D)	F	S	

7. Discussion

We discuss several issues related to applying the reinforcement learning to real robot tasks.

(a) comparison with conventional approaches

In order to compare the performance of our method with conventional approaches, we first implemented a servo mechanism with a controller for angular and forward motion velocities (ω, v). The following equation shows the relationship between (v, ω) and two angular velocities (ω_r, ω_l) to be sent to the right and left motors.

$$\begin{pmatrix} v \\ \omega \end{pmatrix} = \begin{pmatrix} \frac{R_r}{2} & \frac{R_l}{2} \\ \frac{R_r}{T} & -\frac{R_l}{T} \end{pmatrix} \begin{pmatrix} \omega_r \\ \omega_l \end{pmatrix} \tag{9}$$

where R_r, R_l, and T denote the radii of the right and left wheels, and the distance between two wheels, respectively.

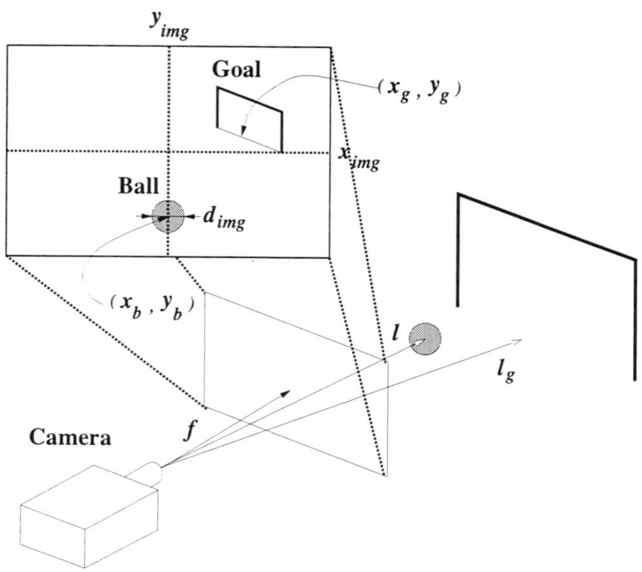

Figure 15. Relationship between camera, ball, and goal

We first implemented the following proportional controller which tries to reduce the differences in visual angle between the lines of sight to the ball center (x_b) and to the goal center (x_g), and in distance between the robot and the ball (l):

$$\omega = k_\theta(-x_b + x_g)/f, \quad \text{and} \quad v = k_l(l - l_d),$$

where f, x_b, x_g, l indicate the focal length of camera, the horizontal coordinate of the ball center in the image, the horizontal coordinate of the goal center, and the distance from the camera to the goal center, respectively, as shown in Figure 15. l_d shows the desired value of the distance between the ball and the robot, and therefore it is zero.

With only this controller, the success rate is worse than the learning method. Then, we add the following controller when the ball is observed between two goal posts:

$$\omega = \dot{\theta} + k_\theta(\theta - \theta_d), \quad v = \dot{l} + k_l(l - l_d), \quad \text{and} \quad \theta = -x_b/f,$$

where θ_d denotes the desired value of the line of sight viewed from the robot. Table 2 compares the performances by the servo control ($k_\theta = k_l = 0.3$) and our learning method. The success rate by the servo mechanism with two controllers is better than that by our learning method because the servo control mechanism assumed *a priori* knowledge on the external and internal parameters of the camera and the robot such as the focal length of camera, the distance to the ball, the distance between two wheels, and so on. However, the

Table 2. Comaprison with servo control

	Servo Control with one controller	Servo Control with two controllers	Learning Method
Success Rate	72.3%	98.6%	78.5%
Mean Steps	64.3	73.2	56.6

mean steps necessary for getting a shoot by the two servo controls are longer than that by the learning method. Because, the servo control tries to turn its body so that the ball and the goal can be observed in a line, and therefore the trajectory of the robot are oscillating a little bit. While, the trajectory by the learning method does not show such a trajectory. It seems possible to obtain the straight path by the servo control method by selecting appropriate gains k_θ and k_l. However, this needs much more knowledge on dynamics of the robot and other parameters such the so-called "viscous friction."

Next, we tested Fuzzy rules which were programmed by a student who has observed the robot behavior by computer simulation for a long time. The rules are as follows:

```
IF ball-is-observed
    IF ball-size-is-large
        IF goal-position-is-center
            forward;
        ELSE IF goal-position-is-left
            right-backward;
        ELSE IF goal-position-is-right
            left-backward;
        ELSE                                   /*goal-is-not-observed*/
            IF goal-is-lost-into-left
                right-backward;
            ELSE                               /*goal-is-lost-into-right*/
                left-backward;
    ELSE                                       /*approach-to-ball*/
        forward;
ELSE                                           /*ball-is-not-observed*/
    IF ball-is-lost-into-left
        left-turn;
    ELSE                                       /*ball-is-lost-into-right*/
        right-turn;
```

We tested the above program with various kinds of conditions in terms of the maximum velocity of the robot and the rolling velocity of the ball (quick, normal, and slow). The success rates by the above program and the learning method are not so different from each other. However, the mean steps to the goal state by the above program is about 10% shorter than that by the learning method. The main reason seems that in the learning method the robot might have experienced useless trials such as kicking a ball by backward motion, which causes inappropriate updates of Q-values in some states, and therefore the optimal path obtained by the learning method might include detours. Note again, the above program is carefully designed by a programmer.

Although the learning method is not superior to the servo control in success rate, or to the Fuzzy rules in mean steps to the goal state, the performance by the learning method is encouraging since the difference does not seem large in spite of not requiring human intervention.

(b) state space construction

The "state-action deviation" problem can be considered as one form of the "perceptual aliasing problem" (Whitehead & Ballard, 1990). It is generally defined as "a problem caused by multiple projections of different actual situations into one observed state". The multiple projections make it very difficult for a robot to take an optimal action. There are several causes for it:

1. Sensor noise: physical sensors are not free from noise, because of which one to many and/or many to one correspondences between actual situations and observed states occur. In our case, the edges of the ball and/or the goal were often not detected, and phantom edges were sometimes detected. Due to such failures, incorrect mapping from one observation to a mismatched state occurs.

2. Sensor processing delay: sensor processing needs non-negligible time to take an adequate action for the observed state. This means that what the robot perceives (state discrimination) is what the environment was some time before. In our case, we have a delay of at least two video frames: one is for image acquisition and another for image processing. We compared the performance of the learned policies with and without delay assuming that the real robot needs two video frames time (66ms). The success rates are 75% by the policy without delay and 60% by the policy with delay. Therefore, we used the learned policy with no delay on a real robot.

3. Difficulty in constructing the consistent staqte and action spaces: as mentioned above, the state and action space are not discrete but continuous in the real world, therefore it is difficult to construct state and action spaces where one action always corresponds to one state transition. We called this the **"state-action deviation problem"**. In utilizing the vision sensor(s), the state-action deviation problem occurs due to the peculiarities of visual information. If we try to construct the state and action spaces in which one action always corresponds to one state transition, we need a large state space that is almost continuous, and suffer from very large variances in the state transition probabilities. Both of these are highly undesirable for the learning method to converge to the optimal solution. We first constructed the state and action spaces which naturally segment the

observed state and the effect of the actuator(s), respectively. These coarse segmentation of the state space ("right" or "left" and "small" or "large") is useful in two ways: by reducing the size of the state space and by absorbing the image processing noise. Then, we reconstructed the action space in such a way that one action consists of a sequence of the same action primitive which is successively executed until the current state changes. This also contributed to reducing the delay in sensor information processing.

4. Hidden states: if there are hidden states which cannot be physically observed by the robot sensors, the system cannot cope with these states without any *a priori* knowledge on them. In our case, the robot easily looses the ball because of its narrow visual field (only 36 degs.). The state of "ball-lost" has a broad area in the real world, and therefore it is difficult for the robot to take an adequate action for this state. Finally, we separated the "ball-lost" state into two states by memorizing the previous state: "ball-lost into right" or "ball-lost into left." This helps the robot take an adequate action to find the ball.

(c) LEM

The LEM algorithm differs in some aspects from the existing approaches in speeding up the search time. In the task decomposition approach (Connel & Mahadevan, 1993), Q-learning is closed inside each subtask. In LEM, however, the robot wanders around the field crossing over the states from which it is easy to achieve the goal, even if the robot is initially placed at such states. The LEM just identifies the positions of the easy states. In other words, we do not need to care so much about the segmentation of the state space in order to decompose the whole task, partial knowledge about the degree of difficulty can be used in the LEM method. However it seems difficult to apply such knowledge in a task decomposition approach.

In the Learning with an External Critic (or LEC) (Whitehead, 1991), the robot receives an advise in each state from the external critic. In order to let LEC work correctly, complete knowledge about the evaluation for the action taken in any state is needed. While, only partial knowledge is used in LEM, completeness of the knowledge does not have any effect on the correct convergence of Q-learning. Only the search time is reduced with complete knowledge.

LEM differs from other approaches such as supervised learning (e.g., (Pomerleau, 1993)) and teaching (e.g., (Lin, 1992)) in that the LEM does not need any demonstrations by a teacher while the teaching method needs at least one or more demonstrations on how instances of the target task can be achieved from some initial states, and supervised learning needs much more demonstrations. These demonstrations must show correct ways to the goal in order for the correct convergence of the learning. The LEM only provides the knowledge of the order of the situations to attain the goal. Even though this knowledge is partially incorrect, the convergence of the learning is guaranteed in LEM, while the other two methods seriously depend on the correctness of the knowledge (demonstrations). In other words, the LEM does not teach how to achieve the goal but from where to start.

From the above differences, the LEM is less restrictive and therefore seems more applicable to a wider range of tasks than supervised learning and teaching approaches.

The most serious and important problem is how to construct the state space. If the state space is not partitioned into sub-states, the LEM cannot be applied. Ideally, the agent should construct the state space automatically from the sensory information and experience. Although some results are reported on this problem (Chapman & Kaelbling, 1991, Mahadevan & Connell, 1991), we need much more work to make clear what kind of information is necessary to achieve the goal and how to construct (segment) the state space from it.

(d) Limitations

Since the state space of our learning method consists of only the current situations, and does not reflect the history in the past except the ball-lost states, currently the robot cannot shoot a ball into a goal if the robot is located between the ball and the goal. Task decomposition might be one solution for this problem, that is, find a ball, turn so that the ball and the goal can be observed in a line, and move forward. Another one is to put other references in the filed such as corer posts and field lines. Now, we are investigating on these approaches.

8. Concluding Remarks

We presented a vision-based reinforcement learning method which is, to the best of our knowledge, the first attempt at applying reinforcement learning (Q-learning) to a real robot task with a real-time vision system. We adopted the Learning from Easy Missions algorithm instead of task decomposition in order to speed up the learning time. The state-action deviation problem due to the peculiarity of visual information is pointed out as one form of the perceptual aliasing problem in applying Q-learning to real robot tasks. Thus, we constructed an action space to cope with this problem.

Although the real experiments are encouraging, still there is a gap between the computer simulation and the real system. We have not made the real robot learn but have only executed the optimal policy obtained by the computer simulation. We are planning to make the real robot learn, starting with the policy obtained by computer simulation as the initial value of the action-value function which is to be determined.

Acknowledgments

This research was supported by the Japanese Sience Research Program under the Project Number 06650301 of the Grand-in-Aid for sientific research from the Ministry of Education, Sience, and Culture.

The authors thank three reviewers for their constructive and valuable comments. They also thank Mr. Eiji Uchibe and Mr. Yasutake Takahashi for their efforts in preparing the revised version of the paper.

Notes

1. To the best of our knowledge, only Whitehead and Ballard (Whitehead & Ballard, 1990) used the active vision system and argued the importance of the so-called "perceptual aliasing" problem. However, they have not shown real experiments.
2. A policy f is a mapping from ???S to ???A.
3. Here, we define one trial as one step state transition by one action.
4. An attempt ends when an agent achieves a goal or fails (crosses over the field boundary).

References

Bellman, R. (1957). *Dynamic Programming*. Princeton University Press, Princeton, NJ.

Chapman, D. & Kaelbling, L. P. (1991). "Input generalization in delayed reinforcement learning: An alogorithm and performance comparisons". In *Proc. of IJCAI-91*, pages 726–731.

Connel, J. H. & Mahadevan, S. editors (1993). *Robot Learning*. Kluwer Academic Publishers.

Connel, J. H. & Mahadevan, S. (1993). "Rapid task learning for real robot". In J. H. Connel and S. Mahadevan, editors, *Robot Learning*, chapter 5. Kluwer Academic Publishers.

Fagg, A. H., Lotspeich, D., & Bekey, G. A. (1994). "A reinforcement learning approach to reactive control policy design for autonomous robots". In *Proc. of 1994 IEEE Int. Conf. on Robotics and Automation*, pages 39–44.

Inaba, M. (1993). "Remote-brained robotics: Interfacing ai with real world behaviors". In *Preprints of ISRR'93*, Pitsuburg.

Kaelbling, L. P. (1993). "Learning to achieve goals". In *Proc. of IJCAI-93*, pages 1094–1098.

Lin, Long-Ji (1992). Self-improving reactive agents based on reinforcement learning, planning and teaching. *Machine Learning*, 8:293–321.

Mahadevan, S. & Connell, J. (1991) "Automatic programming of behavior-based robots using reinforcement learning". In *AAAI-'91*, pages 768–773.

Mataric, M. (1994). "Reward functions for accelerated learning". In *Proc. of Conf. on Machine Learning-1994*, pages 181–189, 1994.

Pomerleau, Dean A. (1993). Knowledge-based training of aritificial neural networks for autonomous robot driving. In J. H. Connel and S. Mahadevan, editors, *Robot Learning*, chapter 2. Kluwer Academic Publishers.

Saito, F. & Fukuda, T. (1994). "Learning architecture for real robot systems – extension of connectionist q-learning for continuous robot control domain". In *Proc. of 1994 IEEE Int. Conf. on Robotics and Automation*, pages 27–32.

Sutton, R. S. (1992). "Special issue on reinforcement learning". In R. S. Sutton(Guest), editor, *Machine Learning*, volume 8, pages –. Kluwer Academic Publishers.

Watkins, C. J. C. H. (1989). *Learning from delayed rewards"*. PhD thesis, King's College, University of Cambridge.

Whitehead, S. D. & Ballard, D. H. (1990). "Active perception and reinforcement learning". In *Proc. of Workshop on Machine Learning-1990*, pages 179–188.

Whitehead, S. D. (1991). "A complexity analysis of cooperative mechanisms in reinforcement learning". In *Proc. AAAI-91*, pages 607–613.

Machine Learning, 23, 305–332 (1996)

Learning Concepts from Sensor Data of a Mobile Robot

VOLKER KLINGSPOR klingspor@ls8.informatik.uni-dortmund.de
KATHARINA J. MORIK morik@ls8.informatik.uni-dortmund.de
ANKE D. RIEGER rieger@ls8.informatik.uni-dortmund.de
Univ. Dortmund, Computer Science Dept. LS VIII, 44221 Dortmund, Germany

Editors: Judy A. Franklin, Tom M. Mitchell, and Sebastian Thrun

Abstract. Machine learning can be a most valuable tool for improving the flexibility and efficiency of robot applications. Many approaches to applying machine learning to robotics are known. Some approaches enhance the robot's high-level processing, the planning capabilities. Other approaches enhance the low-level processing, the control of basic actions. In contrast, the approach presented in this paper uses machine learning for enhancing the link between the low-level representations of sensing and action and the high-level representation of planning. The aim is to facilitate the communication between the robot and the human user. A hierarchy of concepts is learned from route records of a mobile robot. Perception and action are combined at every level, i.e., the concepts are perceptually anchored. The relational learning algorithm GRDT has been developed which completely searches in a hypothesis space, that is restricted by rule schemata, which the user defines in terms of grammars.

Keywords: inductive logic programming, robot navigation, combining sensing and action, perceptually anchored concepts

1. Introduction

Robotics offers a variety of learning objectives. We mention some of the well-known applications of machine learning to robotics and contrast them to our approach, before we present our scenario for learning. In order to enhance a robot's high-level processing, planning can be tailored to particular robot tasks (Segre, 1988). Lessons learned from failures of robot actions can be used to enhance the planning capabilities of a robot (Bennett, 1989, Zercher, 1992). Learning is used to link the environment model with the perception when executing a plan (DeJong & Bennett, 1993, Gil, 1994). The plan or the environment model is refined to accommodate practical experience with the plan. These approaches require an almost complete description of the robot's actions and its environment. It is a time consuming task to build up such a description. Moreover, if the robot's task or environment is changing, the description has to be changed accordingly. This makes the application of robots *inflexible*. In our approach, machine learning helps to construct the description of a robot's actions and its environment.

At lower levels of a robot's processing, machine learning can be applied in order to improve its performance (Kaelbling & Rosenschein, 1990, Millán & Torras, 1992, Mitchell & Thrun, 1993, Baroglio *et al.*, 1994). These approaches assume a fully automated mode of robot operation and learning where no human interaction with the robot or the learning component is intended. Therefore, the representation for learning resembles the robot's low-level representation and is *not easily understandable* to human users. It has been argued, however, that human-machine interaction is necessary in order to fully exploit the opportunities of a robot (Nilsson, 1984, Badler *et al.*, 1991). In order to facilitate access to a robot for human users, the low-level representation has to be translated into an understandable form. We address robot applications in which the robot is not completely autonomous but is interacting with human users. In our approach, machine learning is applied to bridge the gap between low-level and high-level representations.

Biologically inspired work in robotics has developed artificial beings that adapt to their environment (Brooks, 1991, Steels, 1993). This type of learning is restricted to reflex-like behavior. Higher levels of cognition such as reasoning and concept formation are excluded. In contrast, we are interested in the link between perception and concepts. This is the problem of *symbol grounding* (Harnad, 1990, Wrobel, 1991). In our approach, machine learning characterizes concepts on the basis of perceptual and action features, i.e., the learned concepts are perceptually anchored.

Up to now, if a robot has to navigate in office rooms, the coordinates of the walls and doors, desks and cupboards are entered into the robot as a map – for each new set of rooms anew. However, on an abstract level, offices are looking more or less the same wherever they are. Moving in an office environment can be described by concepts such as:

move through the doorway, turn left, move until the cupboard, and stop.

These concepts combine sensing and action. They do not describe a door independent of the robot's movement, but a move through a doorway. Similarly, not the cupboard is the concept here, but approaching a cupboard. The turn implies walls to be detected. However, not the walls are the relevant concepts, but the turn with respect to the walls. We have called such concepts *operational concepts*.

Whereas the specific measurements differ in different offices (e.g., doors have different width and depth), the relations between measurements acquired by sensors during the performance of a movement (e.g., moving through a doorway) remain the same. Moving through a doorway means to first perceive the doorposts with the front sensors, then by the sensors at both sides, and then with the rear sensors. Roughly spoken, there is first a time interval in which the front sensors measure decreasing distances, then a time interval in which the sensors at both sides measure a sudden decrease of distances, and finally a time interval, during which the rear sensors measure increasing distances. The relations between sequential sensor measurements, that characterize approaching or passing a doorpost (e.g., a continuous or sudden decrease of distances) remain the same, independent of the particular distance measurements. The particular time intervals depend on the size of the door, the speed of the robot, and the frequency with which sensor measurements are delivered. However, the relations between the time intervals (*first* decreasing distance ahead, *then* . . . , *then* increasing distance at the back) are independent of the particular size of the door. Also the relations between the sensor groups (*in front* is opposite to *in the rear, left* is opposite to *right*) are the same for all moves through a doorway, although the particular side, which is in front when approaching the door, may differ from one move through a doorway to another. Hence, the concepts can be characterized by relations between time intervals, relations between sensor measurements, and relations between sensor groups.

It is our overall learning task to characterize a concept (e.g., moving through a doorway) and distinguish it from other ones (e.g., moving along a wall) on the basis of sensor measurements and robot movements. As our learning goals are concepts that are described in terms of relations, we need a learning method that is capable of handling relational hypotheses. All learning techniques within the framework of *inductive logic programming* (Muggleton, 1992) learn relational concepts. Whereas inductive logic programming has been successfully applied to relatively small training sets, sets of sensor measurements and robot actions become very large, even if only one concept is to be learned. The application of inductive logic programming to sensor data of a mobile robot demands a method of how to cope with such huge data sets. We developed three answers to this demand. First, we did not solve the learning problem in one step, but have split the overall learning task into several learning steps where each learning step uses the results of the previous one. This decreases the size of the training set for each learning step. The lay-out of learning steps is described in Section 2. Second, we designed the representation language (i.e., the particular predicates that are used to describe a concept and those that denote a concept) such that the points in time or time intervals become *deterministic terms* for the rules to be learned. This reduces the complexity of the learning task, as *determinate rules* are easier to learn than non-determinate ones (Džeroski *et al.*, 1992). The maximal arity of predicates, as well as the maximal depth of terms, is restricted. In addition, an upper bound on the length of rules helps to make the learning task tractable. The representation language is described in Section 3. Third, we applied a learning algorithm which exploits a declarative bias: the hypothesis space can be restricted explicitly. The learning algorithm GRDT is described in Section 4. The concept of moving through a doorway and all concepts that are necessary to characterize this concept were learned from sensor data acquired during movements of a mobile robot (the training set). The learning results were evaluated by applying the learned concepts to sensor data acquired during other paths of the mobile robot (the test set). The experiments are described in Section 5.

2. Learning Scenario

The overall learning task of acquiring concepts from sensor data and robot actions can be modeled as learning from examples. Sensor measurements accumulated during robot routes in known environments and classified by a simulation component are the input to learning. We used the data gathered by a mobile robot which moved in a variety of simple rooms. The rooms were completely known and their description was entered into a simulation component. Also the path of the robot was entered. The sensor measurements were then classified by the simulation component, indicating the edge of an object that has most likely been sensed. Of course, in addition to the errors of the sensors misclassifications occur, too. Hence, the training sets always include noisy data.

Concepts, that describe a movement as well as a pattern of sensor data to be sensed during the execution of that movement are the output of learning. The learning results are applied to other route records, the test set. Then we compare the classification of the simulation component with the classification derived by learned rules. This evaluation shows the applicability of machine learning to defining higher-level descriptions of actions,

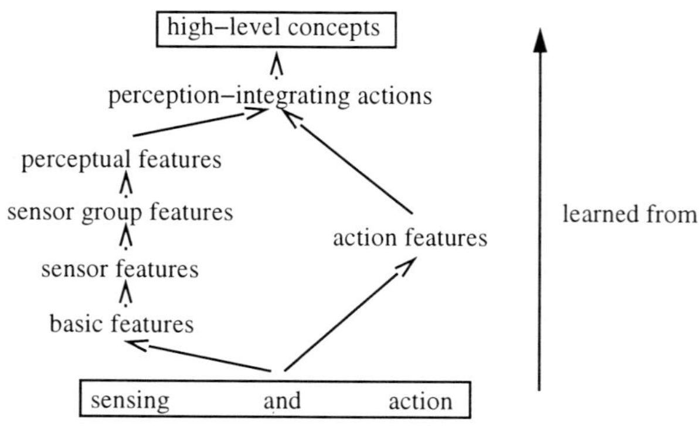

Figure 1. Learning steps

perceptions, and the environment. Moreover, as both, training set and test set, are made of real-world data, it shows how well the learning algorithm can cope with noise and missing measurements. The next step is a real-world test of the learning results. This, however, requires a lot of work that does not concern machine learning (see Section 6 for our ongoing and future work).

Learning high-level concepts directly from sensor data and robot movements is not likely to produce good results. Therefore, we have designed a representation language with intermediate concepts. Five levels of abstraction bridge the gap between the raw data of the robot and the high-level concepts (see Figure 1). Each concept is defined in terms of concepts of the abstraction level below. The learning task is to find these concept definitions.

Now, we can describe the learning tasks. There are several learning steps, each corresponding to a particular level of abstraction. The first step, the learning of basic features from sensor data, is different from all consecutive steps. This learning task corresponds to the problem of signal-to-symbol conversion. At the lowest level, the input is numerical: the robot's movements and sensor measurements are represented by real numbers. The first learning task is then to adjust the calculation of qualitative basic features from the numerical data.

Learning basic features

Given: a sequence of sensor measurements, a set of parameters, and a set of functions for calculating basic features;

Output: the appropriate parameters for calculating basic features and the resulting sequence of basic features.

All following learning steps correspond to the same learning task. The output of one learning step forms the basis for the next one.

Learning concepts

Given: classified instances of a concept and background knowledge;

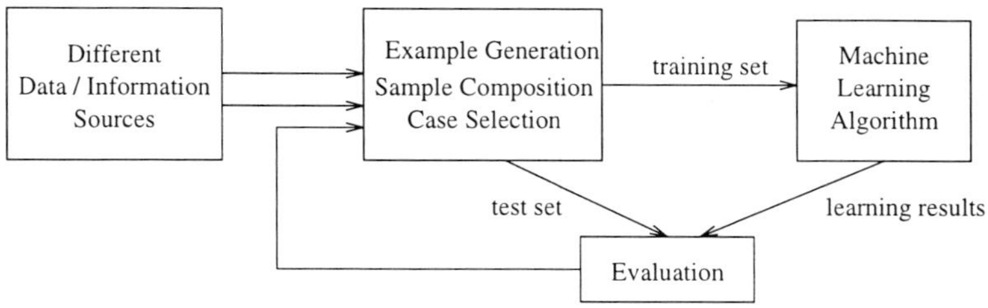

Figure 2. The setting for concept learning.

Output: rules that define a concept in terms of concepts of the next lower level.

A simulation component denotes for every sensor at every time point the edge, that was sensed. This information gives the classification for sensor and sensor group features. Perceptual features are classified by hand. The user of the learning system is supported by a human-computer interface which graphically shows the path of the robot so that the time interval in which sensors should sense, e.g., in_front_of_door, can easily be seen.

The general setting for solving the learning tasks is illustrated in Figure 2. For each learning step, training sets have to be set up. Given the representation, used at each level of the abstraction hierarchy in Figure 1, the respective descriptions have to be derived from the real-world data. Different information sources, such as data files and programs, are used to retrieve and generate predicates, representing classified instances of a concept and background knowledge. This time-consuming example and background knowledge *generation* process is supported and partially automated by a data preparation tool (Rieger, 1995a).

Given a set of positive and negative examples, the user has to decide, which examples are to be included in a training/test set. This decision concerns, e.g., the percentage of positive and negative examples, the construction of several training sets to be used for *cross validation*, and the specification of distributions, according to which examples for a training set are to be drawn. Besides these statistical criteria, the user also makes decisions concerning features of the data, e.g., the data of certain traces may be used for training, data of other traces for testing. Our tool allows the user to specify these decisions for *sample composition* in a uniform framework. With this specification and the given source set of examples, the tool performs the composition automatically.

A training set contains a set of *target predicates*, describing the examples of the concepts to be learned, and a set of *defining predicates*, describing the background knowledge. Target predicates may appear in the conclusion of a learned rule, defining predicates in its premise. Real-word domains, such as robotics, are characterized by an enormous amount of data. In order to apply learning successfully, it is crucial to provide the learning algorithms only with relevant background knowledge. This prevents the algorithm from being overloaded with irrelevant information, which slows down the learning process, and which may even cause the algorithm to find results of poor quality. We provide a method, *case selection*, which allows the user to specify, which defining predicates with which features should be included in the background knowledge for a given set of target predicates. The selection of the relevant defining predicates is then done automatically, yielding *cases*, which associate each target predicate instance with its relevant defining predicate instances.

Thus, the use of our tool makes the data preparation phase more transparent, revealing more clearly its effect on the learning results, and making it amenable to analysis.

3. Representation Hierarchy

High-level concepts integrate perceptual features and action features where the features can be regarded as lower-level concepts. They are anchored in basic features that are learned from sensing and action.

For our scenario, we have chosen the high-level concepts:

move_closer_to_wall, rotate_in_front_of_wall, move_parallel_in_corner, move_along_door, move_through_door, move_in_front_of_door, and rotate_in_front_of_door.

Rules are learned, whose conclusions represent these concepts. The premise of each rule describes the pre-condition for executing a concept, the sensing-acting loop for verifying the concept, and the resulting state (Figure 3).

---

standing(Trace, T1, T2, *in_front_of_door*, PDirect, small_side, *PrevP*) &
parallel_moving(Trace, T2, T3, MSpeed, PDirect, *through_door*, right_and_left) &
standing(Trace, T3, T4, *in_front_of_door*, rear, small_side, *through_door*)
→ move_through_door(Trace, T1, T4).

Figure 3. A high-level concept.[1]

The rule describes the high-level concept of moving through a doorway. It combines the recognition of the doorway with the action of moving through it on a parallel track. Moving diagonally through a doorway is a different concept. Action and perception are linked by the perceptual features (written in italics) occurring as arguments in the action predicates.

The first premise of the rule states, that the robot is standing in a time interval from T1 to T2 of a particular action sequence (Trace) and senses the perceptual feature in_front_of_door with the sensors of a small side after having perceived the perceptual feature *PrevP* prior to T1. This premise denotes the precondition for moving parallelly through a doorway.

The second premise states, that in the following time interval the action parallel_moving is executed, which is associated with the robot measuring the perceptual feature through_door with the sensors at its right and left side. Note, that the particular values for the time intervals do not matter for the concept but are instantiated by the specifics of a particular path. Only the time relation is important. This makes the concept independent of the particular depth of a doorway and the robot's particular rate of advance.

The third premise describes the end of the movement through the doorway: the robot is standing in front of the door, now sensing with its rear sensors mounted at its small side the feature in_front_of_door after having perceived the feature through_door. PDirect, the orientation of perception (e.g., left, right, front, rear) is not fixed by the rule. It is only relevant that the orientation should not change during the parallel move and that, with respect to this orientation, the doorway is sensed in the rear after the movement. That means, whatever the orientation was at the beginning of the action, it is defined as being front, so that after completing the action the doorway is sensed by the rear sensors.

The predicates used in the concept are *perception-integrating action features*. This is the level below high-level concepts in Figure 1. The predicates are standing, moving, parallel_moving, diagonal_moving and rotating. The parallel_moving-predicate (Figure 4), e.g., states that the robot moved in direction MDirec starting at time T1 and it perceived Perc throughout the duration of that move and kept moving until the perception changes at time T2.

```
move(Tr, T1, T2, Speed, MDirec) &
period_of_time_perception(Tr, T1, T2, Perc, PDirect, PSide, parallel)
→ parallel_moving(Tr, T1, T2, Speed, MDirec, Perc, PDirect).
```

Figure 4. A perception-integrating action feature.

Some arguments of a high-level concept refer to *perceptual features*. These are defined by a hierarchy of rules. We may trace the feature through_door from its use as an argument in the concept move_through_door to its basic perceptual features.

```
sg_jump(Tr, right_side, T1, T2, parallel) &
sg_jump(Tr, left_side, T1, T2, parallel) &
Start ≤ T1 & T2 ≤ End
→ through_door (Tr, Start, End, parallel).

sg_jump(Tr, right_side, T1, T2, diagonal) &
sg_jump(Tr, left_side, T3, T4, diagonal) &
succ(T1, T3) & Start ≤ T1 & T2 ≤ End
→ through_door (Tr, Start, End, Move).
```

Figure 5. A perceptual feature.

A set of learned rules determines the meaning of the perceptual feature through_door (see Figure 5).[2] Other predicates are: along_door, in_front_of_door, along_wall, in_front_of_wall, and parallel_in_corner. The second rule accounts for the situation that the door post is perceived by the left hand side sensors a little later than by the right hand side ones. This is expressed by the successor relation between T1 and T3, the starting times of the sensor group feature sg_jump.

Sensor group features and *sensor features* describe configurations of edges as they are perceived by (a group of) sensors during a certain movement. Where, at higher levels of the representation hierarchy, concepts are denoted (e.g., wall, door, corner), here we refer to edges. Configurations of edges are jump, line, concave, convex.

In Figure 6, the edges 1 and 3 as well as edges 4 and 6 are in the relation jump; edges 1 and 2 as well as 4 and 5 are in the relation convex.

If several sensors at the same side or corner of the robot perceive the same edge configuration, the feature predicate has the prefix sg_, denoting a sensor group feature. In Figure 7,

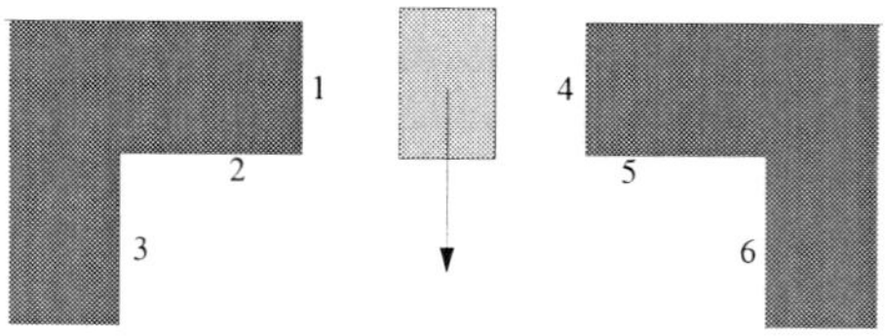

Figure 6. Edge configurations.

in order for the sg_jump feature to hold, three of five sensors at the right side of the robot must perceive the jump configuration when moving parallelly along the edges.

---

s_jump(Tr, S1, T1, T2, parallel) & sclass(Tr, S1, T0, T7, right_side) &
s_jump(Tr, S2, T3, T4, parallel) & sclass(Tr, S2, T0, T7, right_side) &
s_jump(Tr, S3, T5, T6, parallel) & sclass(Tr, S3, T0, T7, right_side) &
succ(T1, T3) & succ(T3, T5) & $T0 \leq T1$ & $T6 \leq T7$
→ sg_jump(Tr, right_side, T1, T6, parallel).

Figure 7. A sensor group feature.

The sensor feature s_jump for a single sensor is defined in terms of basic perceptual features. From sensor measurements we conclude edge groupings.

---

stable(Tr, Orien, S, T1, T2, Grad1) &
incr_peak(Tr, Orien, S, T2, T3, Grad2) &
stable(Tr, Orien, S, T3, T4, Grad3)
→ s_jump(Tr, S, T1, T4, parallel).

Figure 8. A single sensor feature.

The rule in Figure 8 states that a sensor S with a particular orientation Orien measures a sequence of distance values, that have during the time interval from T1 to T2 the gradient Grad1, from T2 to T3 the gradient Grad2, and from T3 to T4 the gradient Grad3. The end point of one time interval is the beginning of the next. An interval with a gradient approximately equal to zero is labeled stable. Grad2 is a rapid increase of measured values denoted by the basic feature incr_peak. stable and incr_peak are basic features that are incrementally calculated from the incoming sensor data (see Section 4).

Note, that the perceptual features are action-oriented. The type of movement with respect to the sensed object (e.g., parallel) is carried along the various levels of perceptual features. In this way, action and perception are integrated not only at the highest level, but also at all lower levels. Except for the basic actions, no feature refers to action or perception alone. *Basic actions* describe motions of the robot during a specific time interval. Possible predicates for actions are move, rotate, and stand, further specified by speed and direction.

measurement (Trace, Time, Sensor, Distance, Sx, Sy, SAngle, Object, Edge).
measurement (t35, 45, s2, 4.96, 4.84, 3.43, 353.54, 0, 5).
measurement (t35, 46, s2, 4.88, 4.92, 3.37, 353.54, 0, 5).
decreasing (t35, 353, s2, 45, 78, -38).

Figure 9. Measurements and a corresponding basic feature

The representation formalism is a restricted Horn logic. This formalism allows an elegant handling of time, time intervals, and relations between time intervals. The relational representation of time intervals makes concept definitions applicable to a variety of actual measurements. The definition of move_through_door applies to doorways of varying breadth and depth. No precise distances need to be fixed by the definition. Similarly, relations between sensors of a sensor group can easily be modeled. The fact that one sensor after the other at one side perceives a corner while the robot is moving along that corner is represented by relations between time instants independent of the particular time distance (see Figure 7).

In addition, the representation formalism allows us to exploit unification. The orientation of perception need not be fixed by the various rules but becomes instantiated when applying a rule (e.g., argument PDirect in Figure 3). The rules explicitly state, that the direction may not change or that another direction must be opposite with respect to the first direction. These abstract relations enable a compact and general representation.

4. Learning Methods

4.1. Learning Basic Features

The calculation of basic features (Wessel, 1995) constitutes the basis for learning. We approach the signal-to-symbol problem by converting numerical sonar sensor measurements to qualitative assertions over time intervals. Basic features are to represent the general tendency of change of the measurements of a sensor during a time interval.

In Figure 10, we see an example sequence of sensor measurements. The measurements of the time intervals from t1 to t4, from t5 to t7, and from t10 to t13 are grouped together, and labeled in accordance with the tendency of change as increasing, decreasing, and (more or less) stable, respectively. During the time interval from t7 to t10 the sensor does not get any echo. This situation is labeled as no_measurement. Between the consecutive time instants t4 and t5, the sensor measurements differ extremely. This situation is labeled as incr_peak. It can be characterized more precisely by considering the *gradient* of the measurements at two consecutive instants of time. It is defined as the quotient, whose numerator is the difference between the measurements at times t+1 and t, and whose denominator is the distance covered by the robot between t and t+1. In the case of incr_peak and decr_peak the gradient is greater than 1 and smaller than -1, respectively. Then, we also know, that two different edges have been perceived at the two instants of time. Therefore, it makes sense to split the time intervals at t4, instead of trying to include t5 in the increasing interval. If

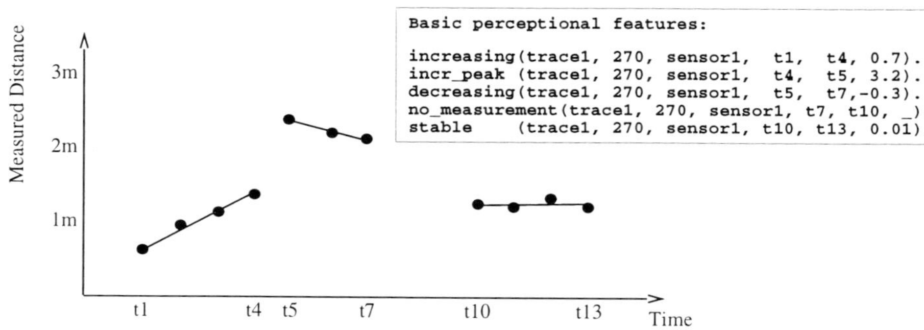

Figure 10. Sensor measurements and corresponding basic features.

the robot does not move during a time interval, we associate the label no_movement with it. If the gradient is approximately equal to -1 or 1, we label the time interval as straight_to and straight_away, respectively.

The calculation of basic features is based on the *incremental* approximation of sensor measurements by (linear) functions. We have not used standard statistical approximation methods, because they don't satisfy our requirement of being applicable in an incremental way during the performance phase [3]. We provide the method with the linear functions, which are used for approximation. As the noisy real-world data cannot be fit exactly to them, we established several parameters, $P_1, \ldots, P_n$, which control the approximation. Each parameter $P_i, 1 \leq i \leq n$ has an associated set of values, $Dom(P_1), \ldots, Dom(P_n)$. We define a *parameter setting* PS as a tuple $(p_1, \ldots, p_n)$, where each $p_i, 1 \leq i \leq n$ represents the value for parameter P_i, i.e., $p_i \in Dom(P_i)$. The parameters, which we consider at the moment, are:

1. *Interval size*, which specifies the maximal number of time instants to be included in a time interval;
2. *Representative gradient* of a time interval. It is either the first or the average gradient of the time interval. It is included as 6th argument in the predicates for basic features (see the examples in Figure 10).
3. *Tolerance* Δ, which is used to compare gradients. A given gradient gr_i is considered approximately equal to the representative gradient gr of a time interval, if $|gr - gr_i| \leq \Delta$.
4. *Alternating sequences*, which are either taken into account or not. If they are not taken into account, an interval is closed, as soon as the current gradient has a different sign than the representative gradient and the tolerance is exceeded.
5. The set of *labels* for time intervals, which constitute the names of the basic feature predicates.

As not all possible parameter value combinations make sense, the set of parameter settings is a subset of the Cartesian product $Dom(P_1) \times \ldots \times Dom(P_n)$. Given the functions (calculating the gradient, calculating the average) and the set of parameter settings, the task is to find the setting, that calculate those basic features, leading to successful learning of concepts.

In the following we present the algorithm for calculating basic features, using a given parameter setting and incrementally reading in one sensor measurement after the other:

1. Initialization: Take the first two measurements, and calculate the gradient.
2. **If**, according to the parameter setting, the new gradient is approximately the same as the previous one
 then include the new time point in the current time interval and adapt the representative gradient as determined by the parameters;
 else close the previous interval at the previous time point and start a new one at the new time point with the new representative gradient being set to the new gradient. Label a closed interval with a symbol, depending on the representative gradient attached to the interval.
3. **If** there are further measurements **then** go to step 2.

We can define a partial order on the parameter settings with respect to the level of detail of the basic perceptional features resulting from the calculation with these parameter settings: A setting PS_1 is smaller or equal than a parameter setting PS_2, $PS_1 \leq PS_2$, if using PS_1 produces more and smaller time intervals than using PS_2. The goal of learning is to find the parameter setting, whose application results in basic features with the appropriate level of detail. In this context "appropriate" means, a sequence of time intervals of basic features corresponds to the time interval of a concept. The learning algorithm can be described by the following steps:

1. Initialization: Choose a member of the set of parameter settings.
2. Calculate the basic features based on this parameter setting.
3. Evaluate the parameter set by comparing the time intervals for basic features and those used by the classification of training instances.
4. **If** the evaluation yields that the parameters produce too few or too many time intervals, **then** choose a parameter setting which is smaller or greater according to the order defined above and go to step 2.

Ultimately, the quality of basic features is determinated by success or failure of concept learning. It is impossible to learn good concept descriptions, if the time intervals determined by the basic features do not match the time intervals used in the classification of examples (see Section 5.1).

4.2. Learning Concepts

The theoretical background for the task of learning concepts is *inductive logic programming* (ILP), whose aim is the induction of Horn clauses from examples of the concepts and background knowledge. Given background knowledge B (here: ground facts), positive and negative examples $E = E^+ \cup E^-$ (again ground facts), and a hypothesis space H (Horn clauses), the examples have to be consistent with the background knowledge (i.e., $B, E \not\models \Box$) and must not be a consequence of the background knowledge (i.e., $B \not\models E$). In ILP, two different goals for learning are considered. The first one is to find a small set of characteristic rules covering exactly the positive examples and avoiding redundancies (Plotkin, 1971). These algorithms often throw away already covered examples to focus on remaining ones. This setting is called the *definite semantics* of ILP[4]. In contrast, the goal of the *nonmonotonic semantics* is to find all rules covering some positive and no negative examples (Helft, 1989)[5]. Since the rules may cover examples multiply, the set of rules is redundant with respect to the given examples. However, the rules are not logically redundant. Whereas the latter learner usually learns more general descriptions, learning in the definite setting is easier and hence faster, because covered examples must not be further taken into account. Our learning task is to learn many logically non-redundant rules.

The goal of learning as many rules as possible is an important issue in our application. Even on an abstract level there are several ways to move through a doorway or to turn left. Therefore, the concept cannot be captured by a single rule. The variety of concept definitions corresponds to the variety moves through a doorway.

For each of the learning steps, we have to prepare examples E and the background knowledge B. If, for example, we want to learn discriminative rules for deriving the sensor feature s_jump from basic features, E^+ consists of the set of ground facts for the target predicate s_jump and E^- consists of all other time intervals for which s_jump is not true. For this learning step, B consists of the defining predicates, describing relevant basic features, i.e., those features, which refer to the same Trace, Sensor, and time interval. Given the example s_jump(tr10,90,s1,10,18,parallel), the data preparation tool will generate the case with this target predicate instance and the relevant defining predicate instances stable(tr10,90,s1,10,14,_), incr_peak(tr10,90,s1,14,16,_), and stable(tr10,90,s1,16,18,_). In an analogous way cases can be constructed for the other learning steps, yielding the trainings sets for the ILP-algorithms.

4.3. GRDT

In principal, most of the ILP-algorithms can deal with our learning tasks. Since the hypothesis space for learning in Horn logic is huge — even if we abstain from proper functions — the different algorithms have different ways to restrict the search.

Semantically pruning the search restricts the hypothesis space dependent on the given data, leading the search into areas of special interest and pruning parts that seem to be unnecessary to characterize the goal concept. The *information gain heuristic*, used by FOIL (Quinlan, 1990), is an example for a semantic restriction.

Syntactic bias, particularly language bias, is used by many different systems to restrict the search space (e.g., (Silverstein & Pazzani, 1991, De Raedt, 1992, Bergadano & Gunetti, 1993, Tausend, 1993)). For all of them, the user must be able to provide the structure of the expected rules to generate the search space description. This step depends heavily on the chosen representation language, the signature. Here, we consider two algorithms using language bias, GRENDEL (Cohen, 1994) and RDT[6] (Kietz & Wrobel, 1992). GRENDEL uses *antecedent description grammars*, a variant of context free grammars, to describe the hypothesis space. Instead of usual terminal and nonterminal symbols, terminal and nonterminal literals are used. The terminals of the grammar are the predicates of the domain. During learning, GRENDEL specializes the current hypothesis by extending one of its nonterminals. Then it tests the new hypothesis in a similar, but more foresighted way than FOIL.

RDT uses another way to describe the hypothesis space, namely *rule schemata*. These schemata are second order rules, describing sets of learnable rules of the same structure. Their predicate variables are successively instantiated during learning. After each instantiation, RDT evaluates the possibly partially instantiated hypothesis. Whereas both, FOIL and GRENDEL, learn concepts according to the definite ILP semantic, RDT finds *all* regularities concerning the goal concept. Only if no further specialization of a hypothesis can lead to an acceptable rule, this part of the hypothesis space is pruned in a completeness preserving way.

In this section, we present a *grammar based rule discovery tool*, GRDT, combining ideas of GRENDEL and RDT in order to handle syntactic bias more flexible and more easily. Like GRENDEL, GRDT uses grammars to describe the hypothesis space. But instead of describing the learnable rules themselves, the grammar defines a set of rule schemata whose predicate variables are instantiated during learning. Like RDT, GRDT is a *nonmonotonic* learner.

4.3.1. Grammars defining sets of rule schemata – the hypothesis space

We will now introduce the use of grammars for defining sets of rule schemata which establish the hypothesis space. In a first example, rules are to be learned where an object must have two specific properties to derive the goal concept. For this task the following rule schema can be used[7]:

P (X) & **Q** (X) → **Concl** (X).

RDT tries to find predicate names for the predicate variables **Q** and **P** such that the resulting rule satisfies the acceptance criterion, e.g., the simple rule parent(X) & female(X) → mother(X). This rule schema can be described with a single GRDT grammar rule:

Concl (X) ⇝ **P** (X), **Q** (X).

Whereas both, RDT and GRDT, determine the predicate symbols to be used during runtime, GRENDEL needs the predicate names occurring in the grammar:

body (mother (X)) ↪ *predicate* (X), *predicate* (X).
predicate (X) ↪ parent (X).
predicate (X) ↪ female (X).
predicate (X) ↪ male (X).
⋮

This is a disadvantage for the user, who, dependent on the domain, may have to specify numerous predicates.

If the number of necessary properties of an object X should not be fixed, rule schemata of different length are necessary:

$\mathbf{P}_1$ (X) → **Concl** (X).
$\mathbf{P}_1$ (X) & $\mathbf{P}_2$ (X) → **Concl** (X).
$\mathbf{P}_1$ (X) & $\mathbf{P}_2$ (X) & $\mathbf{P}_3$ (X) → **Concl** (X).
⋮

The maximal necessary length of the rule schemata must be known a priori to learn all expected rules. RDT orders the different rule schemata according to an extended θ-subsumption (Kietz & Wrobel, 1992, Plotkin, 1970). If a specific instantiation of a general rule schema (e.g., grandparent (X) → mother (X)) covers too few positive examples, i.e., this hypothesis is too special, every more special rule schema (e.g. grandparent (X) & $\mathbf{P}_2$ (X) → mother (X)) will be pruned, because it cannot cover more positive examples than the more general hypotheses do. While the ordering of sets of hypotheses and its use for pruning is considered an advantage, the specification of premises for all lengths is a disadvantage for the user of the learning tool. In contrast to RDT, GRDT grammars allow to define *all* possible schemata with three simple grammar rules:[8]

Concl (X) ⇝ *it* (X).
it (X) ⇝ **P** (X).
it (X) ⇝ **P** (X), *it* (X).

GRENDEL's grammar is similar to GRDT's, but, again, the predicate names have to occur in the grammar:

body (goal (X)) ↪ *it* (X).
it (X) ↪ *predicate* (X).
it (X) ↪ *predicate* (X), *it* (X).
predicate (X) ↪ ...

Let us now consider rules to be learned where a predicate occurs multiple times as a premise of a rule with different arguments:

P (X_1, X_2) → **Concl** (X_1, X_2).
P (X_1, X_2) & **P** (X_2, X_3) → **Concl** (X_1, X_3).
P (X_1, X_2) & **P** (X_2, X_3) & **P** (X_3, X_4) → **Concl** (X_1, X_4).
⋮

With these schemata, for instance, we are capable to learn a rule characterizing ancestor-lines of length three: ancestor(V1,V2) & ancestor(V2,V3) & ancestor(V3,V4) → ancestor_line_3(V1,V4). Unfortunately, the different rule schemata cannot be ordered along θ-subsumption, because of the binding of the head-variables. So, instantiation and testing starts from scratch with each new schema. GRDT uses another way for pruning. Usually, each expansion of a nonterminal specializes the current hypothesis, because the hypothesis cannot become shorter by this step. Then, all further expansions of a hypothesis can be dropped, if the current hypothesis is already too special.

In GRENDEL, for each transitive predicate, a subgrammar must be created. If, for instance, we have the two transitive predicates male_ancestor and female_ancestor, two subgrammars must be given.

$$
\begin{array}{ll}
\mathsf{body}\ (\mathsf{goal}\ (X_s, X_e)) & \hookrightarrow it\ (X_s, X_e). \\
it\ (X_p, X_e) & \hookrightarrow it\_male\_ancestor\ (X_p, X_e). \\
it\ (X_p, X_e) & \hookrightarrow it\_female\_ancestor\ (X_p, X_e). \\
it\_male\_ancestor\ (X_p, X_e) & \hookrightarrow \mathsf{male\_ancestor}\ (X_p, X_e). \\
it\_male\_ancestor\ (X_p, X_e) & \hookrightarrow \mathsf{male\_ancestor}\ (X_p, X_i),\ it\_male\_ancestor\ (X_i, X_e). \\
it\_female\_ancestor\ (X_p, X_e) & \hookrightarrow \mathsf{female\_ancestor}\ (X_p, X_e). \\
it\_female\_ancestor\ (X_p, X_e) & \hookrightarrow \mathsf{female\_ancestor}\ (X_p, X_i),\ it\_female\_ancestor\ (X_i, X_e).
\end{array}
$$

In contrast, GRDT's grammar allows the user to write predicate variables as arguments of nonterminal literals. Then the predicate variables of the different grammar derivations become identical by unification. The following example shows a grammar defining the previously displayed rule schemata, plus rule schemata of length greater than three premises:

$$
\begin{array}{ll}
\mathbf{Concl}\ (X_s, X_e) & \leadsto it\ (\mathbf{P}, X_s, X_e). \\
it\ (\mathbf{P}, X_p, X_e) & \leadsto \mathbf{P}\ (X_p, X_e). \\
it\ (\mathbf{P}, X_p, X_e) & \leadsto \mathbf{P}\ (X_p, X_i),\ it\ (\mathbf{P}, X_i, X_e).
\end{array}
$$

P can become instantiated by both predicates, male_ancestor and female_ancestor.

Let us summarize the weaknesses of GRENDEL and RDT, that are overcome by GRDT. In GRENDEL, the predicate names themselves occur in the grammar. This results in large grammars, which must be changed whenever the set of predicates of the knowledge base changes. The second problem concerns rules where the same predicate multiply occurs in the premises. For a single predicate it is easy to formulate a grammar defining multiple occurrences. If, however, more than one predicate exists, for which these sequences must be built, the grammar becomes complex: for each sequence a subgrammar has to be specified. Using a grammar for the specification of rule schemata, it is possible to maintain the language bias when the set of predicates changes. The weakness of RDT's rule schemata is their fixed length. With GRDT's grammars rules of increasing length can easily be specified, even if the transitivity of a predicate is to be expressed.

Furthermore, GRDT uses the generalization order given implicitly by the grammar instead of θ-subsumption. In this way multiple testing of the same hypotheses is avoided.

4.3.2. Search in the hypothesis space

The GRDT-algorithm starts with the goal predicate as conclusion and generates the first, the most general hypothesis, depending on the grammar, by choosing the grammar's starting rule and instantiating its left-hand side with the goal predicate. Then, it searches the hypothesis space in depth-first order from general to special hypotheses. Each iteration consists of two steps, the *specialization of the hypothesis* and the *test of the hypothesis*.

Specialization

For specializing a hypothesis four alternatives exist:

- **if** the hypothesis contains an uninstantiated predicate variable, try to find an admissible instantiation of this variable.
  Admissible instantiations are all predicates of the knowledge base with the right arity, right argument sorts, and corresponding to the task structure, defined by MOBAL's predicate structuring tool.
- **else if** the hypothesis contains a nonterminal symbol, try to expand the nonterminal. To expand the nonterminal, the next grammar rule, which has this nonterminal as left-hand side is used to replace the nonterminal with the grammar rule's right-hand side.
- **else if** the hypothesis contains a constant to be learned, try to find an admissible value for this constant.
  GRDT learns constants in exactly the same way as RDT. To learn nominal constants, GRDT tests all remaining instantiations of that variable. For ordered values, the smallest (or largest) term of the instances of this variable is chosen first. If the hypothesis is accepted or rejected, all greater (or smaller) terms, respectively, are pruned.
- **else** no further specialization is possible; start backtracking.

Hypothesis test

After each specialization step, the resulting hypothesis must be tested. The cardinalities of the following sets are counted, again exactly like RDT does:

> positive examples covered by the hypothesis; negative examples covered erroneously; uncovered positive examples; yet unknown facts, that can be derived by the hypothesis (predicted); total number of examples.

Testing a hypothesis means to find all solutions for the goal concept with the hypothesis — which is just a Horn clause — as concept definition and the background knowledge as theory. This test is tractable because the facts are ground (Wrobel, 1987, Morik *et al.*, 1993). Since the knowledge is managed by MOBAL's inference engine IM-2, the test itself is slightly different to a simple PROLOG-call.

The user constructs an acceptance criterion using the cardinalities above. For instance, the user may want a hypothesis to cover more positive examples than it does not cover and a ratio of covered positive examples divided by covered negative examples greater than 2.

The result of counting the items is compared with the acceptance criterion. The evaluation is used for pruning the search safely:

- **if** the hypothesis is too special (i.e., the pruning criterion is satisfied)
  stop specializing and start backtracking;
- **else if** the hypothesis is partial (i.e., not all the predicate variables, constants to learn, and nonterminal symbols of the hypothesis are instantiated),
  continue specialization;
- **else if** the hypothesis is accepted (i.e., the confirmation criterion is satisfied)
  store the rule and start backtracking. Further specialization is not sensible because all more special rules are subsumed by the learned one;
- **else** the hypothesis is too general, but cannot be expanded; start backtracking.

4.3.3. Open Problems

Open problems of GRDT should not be hidden. The specialization step does not guarantee, that the new hypothesis is really more special than the one tested previously, although this is assumed for pruning. Having a grammar defining derivations without any specialization of the derived rules, the termination of the algorithm is not assured. However, if it terminates, the algorithm has carried out a complete search through the hypothesis space. The completeness of GRDT is inherited from the completeness of RDT.

Another problem concerns the way how to exclude multiple search of the same part of the hypothesis space, if different derivations of the grammar yield the same hypotheses.

5. Experiments

Our investigations are conducted in cooperation with the university of Karlsruhe. Their mobile robot PRIAMOS (Dillmann *et al.*, 1993) is used for executing the movements and gathering the sensor data. PRIAMOS is equipped with 24 sonar sensors measuring the distance to the nearest object in the range of the emission cone.

We had data from paths in a variety of simple rooms. The rooms differ in two respects. First, the doors of the rooms differ. They are wider or smaller, have thicker or thinner doorframes and the door is centered or it is near to a wall. Second, we have some obstacles in the rooms, which could easily be muddled with a doorway. 17 of the totally 72 paths are movements through a doorway, and 10 paths are along a door. Each trace is documented by 36 to 118 data sets. In total, we had 82139 measurements, each one consisting of the following entries:

> trace number and point of time; robot position; sensor number, its position and orientation; measured distance; object and edge number of what is sensed.

As mentioned in Section 4, the input to learning consists of sensor data in a known environment. Hence, the sensed point and the object to which it belongs is given. The learning task presented here is to acquire the perceptual feature of moving through a doorway, through_door.

For learning, we used three different training sets, each consisting of 26 of the 72 traces and about 2500 measurements. In two of the training sets, the robot moved 11 times, in a third training set, it moved 13 times through the doorway. For verifying the learned rules, we used the remaining 46 traces as test set, containing those examples of movements through a doorway, which were not in the respective training set (6 and 4 examples, respectively). In addition, the test sets contained 11 situations looking similar to movements through a doorway.

We now describe the results of learning three types of rules:

- Rules for patterns for single sensors (see Section 5.1)
- Rules for patterns for groups of sensors (see Section 5.2)
- Rules for moving through a doorway in terms of patterns for sensor groups (see Section 5.3)

For each type of rule, background knowledge, examples, and a grammar defining a set of rule schemata is prepared. We illustrate the learning step by step in the next sections.

5.1. *Learning rules for single sensors*

In this section, we describe how to learn rules for single sensor features. To test the influence of the choice of basic features (Section 4.1), we used three different ways of calculating basic features of different granularity.

The first variant has the least refined granularity. Basic features are intervals of a fixed length, the symbol that is attached to each interval depends on the gradient between the first two measured distances. Totally, a set of 7693 basic features was constructed.

The second variant, with a medium granularity, calculates the length of the sequence dependent on the average gradients between each pair of consecutive measurements. With this variant, we got a set of 9842 basic features.

For the variant with the finest granularity, we additionally regard the divergence of the measurements and add the new measurement to the interval only if the divergence is below a threshold. Now, we calculated a set of 14261 features.

The first learning step was to find sequences of basic features characterizing sensor features, i.e., a specific constellation of edges sensed by a single sensor. For describing the hypothesis space, the following grammar is used:

Concl (Tr, Se, T_1, T_2, **Movement**) $\leadsto$ *seq* (Tr, Se, T_1, T_2, **Movement**).
seq (Tr, Se, T_i, T_2, **Movement**) $\leadsto$ **BF** (Tr, _, Se, T_i, T_2, _).
seq (Tr, Se, T_i, T_2, **Movement**) $\leadsto$ **BF** (Tr, _, Se, T_i, T_j, _),
seq (Tr, Se, T_j, T_2, **Movement**).

GRDT requires the definition of an acceptance criterion. We have chosen a criterion accepting all hypotheses covering more positive examples than they derive additionally ($pos \geq pred$). For four different goal concepts (namely s_line, s_jump, s_convex, s_concave) we used the different training sets for learning resulting in different sets of rules. In Table 1, we have summarized the results of learning rules for single sensors. It shows for each basic feature set the average number of learned rules, counted for the different goal concepts separately. The next column shows the total number of positive test instances for the four concepts. The third value is the total number of instances, derived by the learned rules. From all derived instances some are derived correctly (next column), i.e., they are instances of the concept. The last two columns present the coverage and the correctness.

The results clearly indicate the choice of the basic features to be crucial. Tracing learning explains, why the first set of basic features produces poor results. The time intervals of the basic features do not often fit the time points of the examples, i.e., the sensor features. So only few of the given examples could be used for learning, resulting in a very low number of correctly derived instances. The second basic feature set seems to have the best granularity among the chosen sets.

In addition to the tests described previously, we tested RDT, FOIL, and GRENDEL. RDT learned exactly the same rules as GRDT did, because the hypothesis spaces are equal.

Table 1. Learning results for sensor features.

	#rules	#given	#derived	#correctly der.	correct/given	correct/derived
Basic Feature Set 1						
GRDT	1.9	5113	541	279	5.5%	51.6%
Basic Feature Set 2						
GRDT	12.8	6238	3278	1750	28.1%	53.4%
GRENDEL	3.5	3252	1527	329	10.1%	21.6%
Basic Feature Set 3						
GRDT	7.6	6238	5543	589	9.4%	10.6%

However, because of the different pruning techniques, GRDT is up to five times faster than RDT.

To run FOIL, we prepared negative examples explicitly, because otherwise, FOIL runs out of memory, even if we restrict the automatic generation of negative examples to 1%. We provided all sensor features, that are not currently goal features as negative example. For instance, if s_jump is the goal predicate and s_concave (trace3, sensor1, 4, 19, parallel) is an instance of s_concave, we added not (s_jump (trace3, sensor1, 4, 19, parallel)). From this data, FOIL learned a single, very unintuitive rule:

no_movement (Tr, Or, Se, T_1, T_2, G) $\rightarrow$ s_jump (Tr_2, Se, T_3, T_4, parallel).

This rule is useless, because sensor features describe perceptions while the robot is moving.

The third, and most interesting, comparison is the one with GRENDEL. For this test, we have taken the best basic feature set, Set 2, and a grammar corresponding to GRDT's one. Table 1 summarizes the results, which clearly indicate that for our domain, a learner that just covers the goal concept learns too few rules, that are unlikely to be applicable in future scenes.[9]

5.2. *Learning rules for groups of sensors*

In the next step, we learned rules defining features for groups of sensors. Here, we used grammars focusing the search to hypotheses which constrain the number of sensors of a specific sensor class perceiving the same single sensor feature. We considered classes with one, three, and five sensors. Via the grammars, we also constrained the time intervals during which the pattern must be perceived by the sensors. The largest difference between instants of time is defined by the background knowledge predicates d$i$succ, $i = i, \ldots, 5$. Some rules of the grammar used for learning rules for groups of two sensors, belonging to a sensor class with three sensors, are

Concl (Tr, SCl, T_1, T_2, **M**) $\leadsto$ *sf_comb* (Tr, T_1, T_2, **M**).

sf_comb (Tr, T_1, T_2, **M**) $\leadsto$ **SF** (Tr, Se_1, T_1, T_2, **M**),

Table 2. Learning results for sensor group features.

	#rules	#given	#derived	#correctly der.	correct/given	correct/derived
CC1	193	23000	20642	19553	85.0%	94.7%
CC2	152	30948	31137	26436	85.4%	84.9%

SF (Tr, Se_2, T_1, T_2, **M**),
sclass (Tr, Se_1, First, Last, **SCl**),
sclass (Tr, Se_2, First, Last, **SCl**),
First $\leq T_1$, $T_2 \leq$ Last,
$Se_1 < Se_2$.

sf_comb (Tr, T_1, T_4, **M**) $\leadsto$ **SF** (Tr, Se_1, T_1, T_2, **M**),
SF (Tr, Se_2, T_3, T_4, **M**),
sclass (Tr, Se_1, First, Last, **SCl**),
sclass (Tr, Se_2, First, Last, **SCl**),
First $\leq T_1$, $T_4 \leq$ Last,
Succ(T_1, T_3).

For learning rules describing sensor group features (i.e., sg_line, sg_jump, sg_convex, and sg_concave) in terms of sensor features, we built learn and test sets from the same traces as in the first learning step. We have tested two different acceptance criteria, the first (CC1) accepts all rules covering at least one positive example. The second (CC2) is the same as for learning sensor features, accepting rules if they cover more examples than they predict. Table 2 shows the result of learning evaluated with the test set. The columns represent the same figures as in the previous table. The correctness of the learned rules surpasses that of the first task. The number of learned rules is very high, resulting from the high number of chosen sensor classes. Because of current experiments, we are optimistic to learn the *necessary* sensor classes using conceptual clustering.

5.3. *Learning perceptual features of moving through a doorway*

The last learning step is to combine features sensed by different sensor groups of the robot to describe a perceptual feature like moving through a doorway. The quality of the rules learned in this step is the most relevant one, because it determines whether the whole set of rules can be applied successfully in different environments or not.

For learning, we prepared three training sets, each with 21 traces. In 13 of these traces, the robot moved through the doorway of the room, 6 times parallelly to the doorframes and 7 times diagonally. The hypothesis space is defined by a grammar specifying only a single rule schema:

Concl (Tr, T_{start}, T_{end}, **Movement**) $\leadsto$ **SGF** (Tr, **SClass**$_1$, T_1, T_2, **Movement**),

Table 3. Results of applying all learned rules

	#rules	#given	#derived	#correctly der.	correct/given	correct/derived
Tr. Set 1	12	17	17	12	70.6%	70.6%
Tr. Set 2	11	17	14	11	64.7%	78.6%
Tr. Set 3	10	17	16	12	70.6%	75.0%

$$\mathbf{SGF}\ (\mathsf{Tr}, \mathbf{SClass}_2, \mathsf{T}_1, \mathsf{T}_2, \mathbf{Movement}),$$
$$\mathsf{SClass}_1 \neq \mathsf{SClass}_2,$$
$$\mathsf{T}_{start} < \mathsf{T}_1, \mathsf{T}_2 > \mathsf{T}_{end}.$$

The acceptance criterion accepts hypotheses covering more than 3 positive examples and no negative one.

We used all 72 traces to verify the learned rules. As the test set only contains sensor measurements, learned rules from *all* levels of abstraction were applied to derive the perceptual features. Table 3 shows the results. In every test set, three situations are classified incorrectly, in which the robot moved through the doorway very close to one of the doorposts. Because of the short distance, the sensors did not got any echo and hence, it could not derive the feature through_door.

Overall, the result justifies our approach. The quality of the learned rules improves as the representation hierarchy is climbed. The rules are capable to classify many of the given examples correctly. This way to get more and more abstract concepts is robust against the noise, produced by the weak ultrasonic sensors.

As this test evaluates all rules up to perceptual features, it is the most important one. It shows the advantage of designing a sequence of learning tasks for concepts at more and more abstract levels.

6. Related and future work

We have presented an approach to making a complex overall learning task tractable. The learning of high-level concepts from sensor data gathered during several paths in known environments (particular office rooms) has been split into a sequence of learning steps. The results of all learning steps together form a knowledge base that derives high-level concepts from low-level data. The training phase is off-line. The results of the training phase show that inductive logic programming is capable of constructing high-level descriptions of actions, perceptions, and the environment from sensor data. Although several learning steps – particularly those at lower levels – deliver concepts that misclassify some data, the overall knowledge base achieves a high accuracy for high-level concepts. This robustness encourages us to put the learned concepts to use: concepts are to be executed in unknown environments of a certain type, e.g., office rooms. During the execution phase, the rules characterizing concepts are applied to the actual sensor measurements while performing certain moves. In the next section, we describe how we plan to apply the learned concepts to real-world missions of a mobile robot.

6.1. Future Work

The learned knowledge base is to be used for *task specification, execution monitoring, explanation of error recovery,* and for the *update of the environment representation.* When using the knowledge base for task specification, a planning component is needed that converts high-level concepts to actual perceptions and actions of the robot. The robot user may state a goal concept. The planner[10] then proves the premises of the rule that characterizes the concept and calls the execution of basic actions in the following way:

Given: State (given by an action-oriented perceptual feature), goal concept

Planning:

> **If** the verification of the sensor pattern of the high-level concept has been achieved,
> then stop with the message "success" and output the recordings of actions and perceptions in abstract terms.
>
> **If** no planning for the current state is possible using the given rules
> then stop with the message "fail" and output the recordings of actions and perceptions in abstract terms.
>
> **Else** derive a plan for transferring the current state into a precondition of the goal concept (high-level planning), and
> derive an action with its corresponding perceptual feature (deriving lower-level actions)

Acting: execute the action and acquire sensor data until

- either you are in danger to collide with an object – then change direction
- or you have verified a perceptual feature for an action – then proceed with **planning**.

Action and perception are closely related in this incremental planning strategy. The learned rules are used by backward and forward chaining. Backward chaining creates a plan for transferring the current state into the precondition of the high-level concept. Backward chaining also derives the low-level commands of the robot from the high-level concepts. Forward chaining derives the perceptual features from the corresponding sensor data acquired during a certain move of the robot.

When, for instance, the planner is called with the state

standing(t88, 45, 48, before_wall, right, small_side, along_door)

(standing(*Trace*, *Time*, *Time*, *Perception*, *SensorClass*, *RobotSide*, *PreviousPerception*))

and the goal concept

move_in_front_of_door(t88, 45, *T4*, *T3*),

backward chaining verifies the preconditions of the learned rule for move_in_front_of_door. The first condition is already given by the current state. The second condition (opposite(right,

X)) is verified by matching the fact opposite(right, left), which is part of the knowledge base. The premise

parallel_moving(t88, 48, *T3*, slow, backwards, in_front_of_door, left)

requires further backward chaining in order to achieve basic robot actions. Here, the basic action is move(t88, 48, *T3*, slow, backwards)[11]. This basic action is performed until the sensor pattern in_front_of_door is verified. This is the acting step of the planner. Within this step, forward chaining is used to derive sensor patterns from incoming sensor data. It is for this use of basic features, that their calculation is already performed incrementally. As soon as the sensor group of the robot's left side perceive measurements that indicate the concept in_front_of_door, the action stops and *T3* becomes instantiated with the point in time of the last sensor data necessary to conclude in_front_of_door. The result of the acting step is

parallel_moving(t88, 48, 72, slow, backwards, in_front_of_door, left).

Planning proceeds with proving the verification condition of move_in_front_of_door,

standing(t88, 72, *T4*, in_front_of_door, left, small_side, in_front_of_door).

This evokes again the acting step. For three more points in time the sensors at the robot's left side (which is its small side) measure distances, that are classified as in_front_of_door by forward chaining. The result of the acting step is

standing(t88, 72, 75, in_front_of_door, left, small_side,in_front_of_door).

The goal concept is now verified with the fully instantiated fact

move_in_front_of_door(t88, 45, 75, 72).

Until now, the coupling between the knowledge-base and the robot is "manual", i.e., the data exchange is realized by an engineer who loads the basic actions derived by the knowledge-based system into the robot system and loads the sensor data of the robot into the knowledge-based system. We are currently working on conducting real-world experiments in collaboration with the University of Karlsruhe and their robot PRIAMOS. We are aware of problems with the time constraint of real-time processing that can hardly be met by our planner. This may well force us to distinguish in the future between planning in advance and planning at runtime as do (Gervasio & DeJong, 1994).

We explore the compilation of a rule set and facts into a structure that can more efficiently be executed in real-time. A first approach is to map rule sets to tree automata, whose final states represent concepts, which have to be recognized, in order to verify, that a goal has been achieved. In (Rieger, 1995b), we have shown, how a deterministic finite state automaton for the recognition of sensor features can be inferred. It takes as input a sequence of temporally ordered observations, i.e., basic features. Each of its final states is associated with one or possibly many sensor feature concepts. During the performance phase, we apply a marker passing method to the automaton: When the robot moves through the environment, its sensors constantly perceive observations, each of which might be the beginning of a string, which is accepted by the automaton. For each observation a marker is generated. Each marker is passed, if possible, from its current state to a successor state according to the transition function. If a marker has reached a final state, a message is produced, that an object has been recognized, which is possibly an instance of several concepts.

If a final state is associated with different concepts, we have the situation, that the sequence of observations leading to it, cannot be interpreted unambiguously by the robot. It cannot be sure, in which state it is and has difficulties in choosing the next action. In order to account for these uncertainties arising from ambiguous observations, we infer from the deterministic state automaton a hidden Markov model. This is done by splitting up the ambiguous final states into several unambiguous ones, by introducing non-deterministic transitions to the newly generated states, and by associating probabilities with the non-deterministic transitions. During learning, the training data is evaluated in order to derive relative frequencies, which are taken as estimates for the transition probabilities. During the performance phase, we use a modified version of the *Viterbi algorithm* (Jelinek, 1976), with which we determine the concept, which most probably has been perceived. In this way, the acting phase of our performance step is enhanced.

6.2. Related Work

There are similarities between our work and Shen's approach (Shen, 1993) to learning from the environment. In both cases the objective is to fuse application of rules with their construction, i.e., to couple learning with its evaluation. Shen also applies rule induction from examples in the context of problem-solving. With respect to the bidirectional use, our concepts are similar to the prediction rules used by Shen. The main difference concerns the domain of application, which in our case is much more demanding, as we are dealing with real-world data.

In the JPL Telerobot (Doyle *et al.*, 1986) a STRIPS-like planner is used, in which the precondition lists are amended to contain sensory requests to verify beliefs, and the post-condition lists are amended to include sensory requests that validate the effects of each action. The authors present a representation language in which these sensory requests are to be programmed. The idea being the same, our approach differs in that we represent and learn plans within the ILP-framework.

Our approach to the application of learning to robotics is closely related to approaches to learning plans for robots (Mitchell, 1990), (DeJong & Bennett, 1993), (Gervasio & DeJong, 1994), and (Gil, 1994). Some distinctions need to be mentioned. First, we do not use an a priori model of the environment. Therefore, we do not learn from mismatches between the predicted and the observed effects of actions as (DeJong & Bennett, 1993) and (Gil, 1994) do. Our learning does not refine a given plan, instead it actually constructs the rules to be used for planning. Second, our representation of planning operators is not that of classical planning with add and delete lists which makes inference non-monotonic. Instead, we attach time instants to sensor measurements and, on higher levels, we attach time intervals, so that reasoning remains monotonic. Our rules can be interpreted as operators with preconditions and postconditions, but technically they are just clauses. Third, our representation of states is different from the work cited above. A state corresponds to a time interval in which a particular sensor pattern is valid. The validation of the sensor pattern is left until actual execution. This way, we avoid committing ourselves prematurely to a firm sensor value. Our plan just states that a move has to be executed until a pattern of distance measurements is registered. This is similar to the contingent variables of (Gervasio & DeJong, 1994). Note, however, that in our case the sensor pattern is more complex than just a variable for a particular distance. A perceptual feature in our approach already abstracts beyond particular distance values and gives relations between sensor measurements instead. This abstraction allows a perceptual feature to cover many different situations. The particular time interval is instantiated in the rule by the value determined at runtime and propagated to other rules by unification. Fourth, in contrast to the mentioned planning approaches, our representation combines sensing and action at every level of detail. That is, we do not have separate sensory operators for acting in order to recognize an object, but these actions are treated in the same way as other actions. In our example, it is the along_door feature which recognizes the door so that move_through_door becomes applicable.

7. Conclusions

We have applied machine learning to a challenging task, namely learning a hierarchy of concepts from sensor data and robot actions. We tackled this complex overall learning task by dividing it into several learning steps that link several levels of abstraction. In order to perform learning from examples, we applied a simulation component that classifies measurements with respect to the edges of the known environment that have been sensed. An example generation tool for putting together training sets for one learning step has been developed. A comfortable human-computer interface supports the management of the various training and test sets and shows the training paths graphically.

We have designed a representation language that expresses concepts which fuse perception and action for robot navigation. Even without learning capabilities, the high-level description is useful as has been argued by (Stopp *et al.*, 1994). Their project aims at natural language access to a robot. On one hand, the user can specify the robot's task in a way appropriate for human users. On the other hand, the user can more easily understand the robot's actions. In error situations, user and robot can communicate in order to recover from the failure. We do not go as far in the high-level description as natural language does but consent to Prolog clauses which are already easily understandable for a knowledgeable user. Moreover, natural language processing requires high-level concepts that are perceptually anchored. We learn such concepts.

The application of ILP to sensor and action data goes beyond standard ILP applications. First, the application of symbolic learning to robot data requires the conversion of numerical data to qualitative terms. An algorithm for the calculation of basic features has been developed. Second, the rather large size of the training sets asks for appropriate means to restrict the hypothesis space. GRDT is an ILP-algorithm, which uses grammars in order to define rule schemata that restrict the hypothesis space. Trying to generate negative examples by the closed-world assumption (as FOIL does) leads to tremendously large data sets that cannot be handled. GRDT does not explicitly create a set of negative examples but applies the closed-world assumption implicitly, if the user defines the acceptance criterion such that no new instances of the target concept are to be predicted. Third, as there are several ways for a concept such as, e.g., move through a doorway, to be applied, as many non-redundant rules as possible should be learned. Instead of ending the search for rules as soon as the target concept is covered, GRDT stops the search only, when further hypotheses would be redundant or cannot be accepted (i.e., some of the rule premises already cover too few positive examples).

The learning results have been tested in two ways. First, a test set for the particular learning step has been formed. Here, the learned rules of this step are used independent of other learned rules. Second, the rule base that has been acquired by several learning steps was tested on sequences of basic features. It became clear, that misclassifications at lower levels of abstraction do not result in the misclassification of concepts at higher levels of abstraction. If, for instance, the measurements of one sensor are not right, the sensor group may nevertheless deliver the correct pattern. This is due to (learned) rules that demand three sensors of a group to derive the same pattern before this pattern is said to hold for the sensor group. In this way, the hierarchy of learned concepts contributes to the robustness of learning results. Experiments have shown that GRDT learns concepts from data supplied by a robot effectively and efficiently.

Acknowledgments

The BLearn-II project is funded by the European Community under P7274. The Dortmund part of the project is additionally funded by the Ministry for Science and Research of the Federal State Nordrhein-Westfalen.

We thank Stefan Sklorz, and Stephanie Wessel, who developed parts of the system. In particular, Stephanie Wessel made the system learn basic features. Stefan Sklorz developed the action features and the experimental planner. He also provided the human-computer interface for handling training sets and ran several experiments. Moreover, in the course of many discussions both contributed ideas. Eckart Zitzler does an important job in structuring the access to robot data used by several modules. Michael Goebel has tried out alternatives (e.g., conceptual clustering) for the acquisition of sensor patterns. We also thank Peter Brockhausen for proof-reading earlier versions of this paper. The described system is the result of the collective effort of everybody involved.

We thank wholeheartedly the MOBAL team at the Gesellschaft für Mathematik und Datenverarbeitung (GMD), Werner Emde, Jörg-Uwe Kietz, Edgar Sommer, and Stefan Wrobel. We use the MOBAL-system (Morik *et al.*, 1993) for the maintenance of the knowledge base and parts of GRDT are directly taken from MOBAL's learning component RDT. In particular, Jörg-Uwe Kietz advocated the relational representation of time and assisted us with theoretical effort estimations.

Notes

1. We follow the Prolog convention and write constant terms in small letters, variables start with a capital letter.
2. For technical reasons we had to introduce conversion rules that convert the predicate of a rule's conclusion into an argument of another predicate. An example is the rule through_door (Tr, T1, T2, PDir, Side, Orien) → period_of_time_perception(Tr, T1, T2, through_door, PDir, Side, Orien).
3. Learning an incremental function approximator definitely goes beyond the purpose of our research.
4. This and the following definition are taken from (Muggleton & De Raedt, 1993).

5. Helft's approach actually does not use positive and negative examples but completes the given observations by the generalization step, which implies a closed-world assumption. As we assume our features to be disjoint, all instances of another feature are considered negative examples for the feature currently to be learned. This approach leads to almost the same results as Helft's approach. The difference is, first, the flexible user-given acceptance criterion, applied by GRDT, and, second, the focus on one goal concept for learning in one learning run.
6. RDT is one of the learning tools of the knowledge acquisition system MOBAL (Morik *et al.*, 1993)
7. A short note about our typographical conventions: rule schemata of RDT are written with ordinary arrows ($\rightarrow$), GRENDEL's grammars with $\hookrightarrow$, and GRDT's grammars with $\leadsto$. Nonterminal literals are written *slanted*, terminals are written sans serif. Variables (both, predicate variables and argument variables) that should be learned by RDT or GRDT are written in **bold face**.
8. Variables occurring in different grammar rules are regarded to be different.
9. Because GRENDEL sometimes runs out of memory, too, we did not performed all tests, and hence, the number of given examples is smaller than with GRDT.
10. A small planning component has been implemented by Stefan Sklorz.
11. The sides of the robot are named with respect to the robot's movement such that the direction of the movement always corresponds to the front side. For execution, the relative orientation with respect to the movement is converted to the absolute orientation of the robot, where sensor s0 defines the front side of the robot. In the example here, relative and absolute orientation happen to be the same. When the robot moves backwards with respect to the previous movement, the sides change their names so that they are now named with respect to this movement. In this way, right becomes left when the robot moves backwards. Hence, the door has to be perceived by the sensors at the left side, which before were the sensors at the right side.

References

Badler, N.I., Webber, B.L., Kalita, J. & Esakov, J. (1991). Animation from instructions. In N.I. Badler, B.A. Barsky, and D. Zeltzer, editors, *Making Them Move: Mechanics, Control, and Animation of Articulated Figures*, pages 51–93. Morgan Kaufmann, San Mateo, CA.

Baroglio, C., Giordana, A. & Piola, R. (1994). Learning control functions for industrial robots. In *MLC-COLT '94 Robot Learning Workshop*.

Bennett, S.W. (1989). Learning uncertainty tolerant plans through approximation in complex domains. Technical Report UILU-ENG-89-2204, The Beckman Institute for Advanced Science and Technology, University of Illinois at Urbana-Champaign.

Bergadano F. & Gunetti, D. (1993). An interactive system to learn functional logic programs. In *Procs. 13th IJCAI*, pages 1044–1049. Morgan Kaufmann.

Brooks, R.A. (1991). The role of learning in autonomous robots. In *Procs. COLT 91*, pages 5–10.

Cohen, W.W. (1994). Grammatically biased learning: Learning logic programs using an explicit antecedent description language. *Artificial Intelligence*, 68(2):243–302.

De Raedt, L. (1992). *Interactive Theory Revision: an Inductive Logic Programming Approach*. Academic Press.

DeJong, G. & Bennett, S. (1993). Permissive planning – a machine learning approach to linking internal and external worlds. In *Procs. 11th AAAI*, pages 508–513.

Dillmann, R. Kreuziger, J. & Wallner, F. (1993). PRIAMOS – an experimental platform for reflexive navigation. In Groen, Hirose, and Thorpe, editors, *IAS-3: Intelligent Autonomous Systems*, chapter 18, pages 174–183. IOS Press.

Doyle, R.J., Atkinson, D.J. & Doshi, R.S. (1986). Generating perception requests and expectations to verify the execution of plans. In *Procs. of the AAAI*, pages 81–88.

Džeroski, S., Muggleton, S. & Russell, S. (1992). PAC-learnability of determinate logic programs. *Procs. of 5th COLT*, pages 128–135.

Gervasio, M. & DeJong, G. (1994). An incremental learning approach to completable planning. In W. Cohen and H. Hirsh, editors, *Procs. 11th Int. Machine Learning Conf.*, pages 78–86.

Gil, Y. (1994). Learning by experimentation – incremental refinement of incomplete planning. In W. Cohen and H. Hirsh, editors, *Procs. 11th Int. Machine Learning Conf.*, pages 87–95.

Harnad, S. (1990). The symbol grounding problem. *Physica D*, 42:335–346.

Helft, N. (1989). Induction as nonmonotonic inference. In *Procs. of the 1st Int. Conf. on Knowledge Representation and Reasoning*,.

Jelinek, F. (1976). Continuous speech recognition by statistical methods. *Procs. of the IEEE*, 64:532–556.

Kaelbling, L.P. & Rosenschein, S.J. (1990). Action and planning in embedded agents. *Robotics and Autonomous Systems*, 6:35–48.

Kietz, J. & Wrobel, S. (1992). Controlling the complexity of learning in logic through syntactic and task–oriented models. In Stephen Muggleton, editor, *Inductive Logic Programming*, chapter 16, pages 335–360. Academic Press, London.

Millán, J. & Torras, C. (1992). A reinforcement connectionist approach to robot path finding in non-maze-like environments. *Machine Learning*, 8:363–395.

Mitchell, T.M. & Thrun, S.B. (1993). Explanation-based neural network learning for robot control. In C. L. Giles, S. J. Hanson, and J. D. Cowan, editors, *Advances in Neural Information Processing Systems 5*, San Mateo, CA. Morgan Kaufmann.

Mitchell, T. (1990). Becoming increasingly reactive. In *Procs. 8th AAAI*, pages 1051–1058.

Morik, K., Wrobel, S., Kietz, J.-U. & Emde, W. (1993). *Knowledge Acquisition and Machine Learning - Theory, Methods, and Applications*. Academic Press, London.

Muggleton, S. & De Raedt, L. (1993). Inductive logic programming: Theory and methods. Technical Report CW 178, Department of Computing Science, K.U. Leuven.

Muggleton, S., editor. (1992). *Inductive Logic Programming*. Academic Press.

Nilsson, N.J. (1984). Shakey the robot. Technical Note 323, Artificial Intelligence Center, SRI International, Menlo Park, CA.

Plotkin, G.D. (1970). A note on inductive generalization. In B. Meltzer and D. Michie, editors, *Machine Intelligence*, chapter 8, pages 153–163. American Elsevier.

Plotkin, G.D. (1971). *Automatic methods of inductive inference*. PhD thesis, University of Edinburgh.

Quinlan, J.R. (1990). Learning logical definitions from relations. *Machine Learning*, 5(3):239–266.

Rieger, A. (1995a). Data preparation for inductive learning in robotics. In *Working Notes of the IJCAI-Workshop on Data Engineering for Inductive Learning*. also available as LS8-Report 19, University of Dortmund.

Rieger, A. (1995b). Inferring probabilistic automata from sensor data for robot navigation. In *Procs. of the 3rd European Workshop on Learning Robots*. also available as LS-8 Report 18, University of Dortmund.

Segre, A. (1988). *Machine Learning of Robot Assembly Plans*. Kluwer, Boston.

Shen, W.-M. (1993). Discovery as autonomous learning from the environment. *Machine Learning*, 12:143–165.

Silverstein, G. & Pazzani, M.J. (1991). Relational clichés: Constraining constructive induction during relational learning. In L. Birnbaum and G. Collins, editors, *Machine Learning: Procs. of the 8th Int. Workshop*, pages 298–302, San Mateo, CA. Morgan Kaufmann.

Steels, L. (1993). Building agents out of autonomous behavior systems. In L. Steels and R. Brooks, editors, *The 'artificial life' route to 'artificial intelligence' – Building situated embodied agents*. Lawrence Erlbaum, New Haven.

Stopp, E., Gapp, K.-P., Herzog, G., Laengle, T. & Lueth, T. (1994). Utilizing spatial relations for natural language access to an autonomous mobile robot. In *KI–94, – Procs. of the German Conference on AI*, Berlin. Springer.

Tausend, B. (1993). Representing biases for inductive logic programming. In Pavel Brazdil, editor, *Machine Learning - Procs. of ECML-93*, Lecture Notes in Artificial Intelligence, pages 427–430, Berlin, Heidelberg, New York. Springer.

Wessel, S. (1995). Lernen qualitativer Merkmale aus numerischen Robotersensordaten. Master's thesis, Universität Dortmund.

Wrobel, S. (1987). Higher-order concepts in a tractable knowledge representation. In K. Morik, editor, *GWAI-87 – 11th German Workshop on Artificial Intelligence*, pages 129–138, Berlin, New York, Tokyo. Springer Verlag.

Wrobel, S. (1991). Towards a model of grounded concept formation. In *Procs. 12th IJCAI*, pages 712–719, Los Altos, CA. Morgan Kaufman.

Zercher, K. (1992). *Wissensintensives Lernen für zeitkritische technische Diagnoseaufgaben*. infix, Sankt Augustin, 1992.

Index